Radiation Alert

Other books from Energy Probe

Energy Probe's Statistical Handbook: *The Other Energy Pie*
  (by David Poch)
In The Name of Progress: The Underside of Foreign Aid
  (by Patricia Adams and Lawrence Solomon)
Power at What Cost? (by Lawrence Solomon)
The Conserver Solution (by Lawrence Solomon)
Energy Shock (by Lawrence Solomon)
Over A Barrel (by Jan Marmorek)
Zero Energy Growth for Canada (by David Brooks)

Radiation Alert

A Consumer's Guide to Radiation

DAVID I. POCH

An Energy Probe Project

1985

Doubleday Canada Limited, Toronto, Ontario
Doubleday & Company, Inc., Garden City, New York

ISBN 0-385-19029-8
Library of Congress Catalog Card Number: 83-45160

First edition
Typeset by Compeer Typographic Services Ltd.
Printed and bound in Canada by Gagne Printing Ltd.
Design and illustrations by Annette Tatchell
Cover design and illustration by Annette Tatchell

Canadian Cataloguing in Publication Data

Poch, David I. (David Ira), 1954–
Radiation alert

ISBN 0-385-19029-8

1. Radiation—Safety measures.   I. Title.

RA569.P59 1985   363.1′79   C85-099267-2

Library of Congress Cataloging in Publication Data

Poch, David I. (David Ira), 1954–
  Radiation alert

  Includes index.
  1. Radiation—Safety measures.   2. Consumer
education.   I. Title.

RA569.P59 1985   363.1′79   83-45160
ISBN 0-385-19029-8

Photographs courtesy of Ontario Hydro
Used with permission

For Helen and Lou

Acknowledgments

Many thanks to Jan Marmorek, Irene Clark and Marilyn Aarons for their research, Kate Sutherland for her skillful editing and the many Energy Probe volunteers who helped get the manuscript in and out of the word processor (their exposure to low levels of radiation in the process was greatly appreciated).

Also many thanks to Dr. William Paul, Norman Rubin and to Lawrence Solomon for their comments.

Finally, special thanks to Hinda Goldberg without whose support this project would not have been possible.

Contents

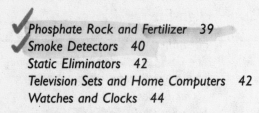

A PRIMER
ON RADIATION

CHAPTER ONE

Radiation:

Just What Is It?

People have a natural fear of the unknown and, in the case of radiation, we are speaking largely of the unknown. Even the experts admit that what we don't know about radiation and its effects could fill volumes.

Before the mid-forties we worried very little about radiation. We wore watches with radium-illuminated dials, and medical X rays were given with abandon.

Then came the dropping of the atomic bomb at Hiroshima followed by bomb testing by the United States, Britain, the USSR and, eventually, China and France. Strontium 90 began to appear in milk and, for the first time, the public began to view radiation as a hazard.

To allay public fears, government P.R. departments produced films with themes like "Duck and Cover" which, arguing that radiation could be treated like shrapnel, taught school children to hide under their desks should an atomic bomb fall. However, when nuclear power plants were built near cities in the 1960s and 1970s, public fears about radiation resurfaced. The P.R. men of the 1970s produced more pamphlets on radiation, this time with the theme "Radiation is part of natural life." They hoped this low-key portrayal would help dispel the public's fears.

The public relations pamphlets are, in a way, correct: We have always been subjected to natural forms of radiation. Long before radiation was ever "manufactured" by household appliances and medical equipment, it was emitted naturally by radioactive substances—such as uranium. But the fact that radiation occurs naturally does not make it less dangerous—arsenic, hemlock and many other poisons also occur naturally. And as far as our health is concerned, the source of the radiation is less important than the type and amount we receive. Whether it is natural or unnatural, the effects of radiation are the same.

Radiation only now is beginning to obtain the respect it deserves as both a natural and a man-made opponent. Although we still have lots to learn about it, what we do know is enough to warrant strict prudence about its use. Phrases describing radiation as having "no risk" and as being "nothing to worry about" are no longer applicable.

High-Energy *versus* Low-Energy Radiation

Although radiation takes a variety of forms, all have several features in common including the simple fact that radiation is a form of energy.

You have probably already heard of many of the common forms of radiation, among them:

- radio waves, which carry sound to transistor radios;
- microwaves, which cook food;
- infrared light, which comes from all warm objects;
- visible light, which we rely on to see;
- ultraviolet light, which gives us suntans;
- X rays, which allow doctors to photograph our insides;
- gamma rays, which are used in radiation therapy; and
- particles known as: alpha particles, beta particles, neutrons, protons and positrons, several of which together with gamma rays are emitted from radioactive substances used in nuclear power plants.

We can detect only two of these with our senses: visible light can be seen and infrared radiation can be felt as heat. All other forms are undetectable to us without the help of precision instruments.

Each form of radiation has its own level of energy. Low-energy forms include radio waves and microwaves; medium-energy radiation includes infrared, visible and ultraviolet light; and high-energy forms include X rays, gamma rays and particle radiation.

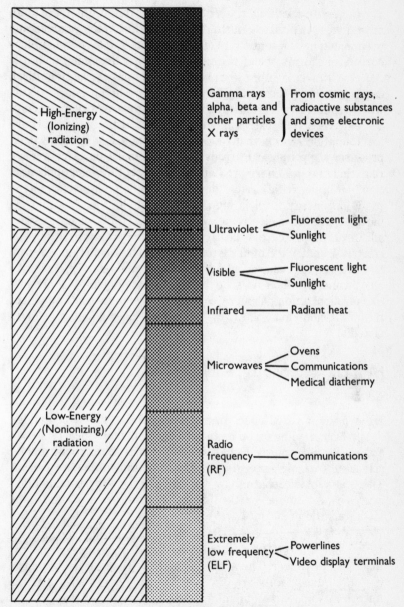

FIGURE I
The Electromagnetic Radiation Spectrum

Generally speaking, the higher the level of radiation, the more harmful it is to human health. But several other factors come into play, including the sensitivity of our bodies and the varying ability of different forms of radiation to penetrate body tissue.

Microwaves are one example of low-energy radiation which can be especially dangerous because they are highly concentrated when produced by various appliances and devices, such as microwave ovens and industrial heaters.

Radiation scientists divide their field of study into two main areas for the purposes of studying biological effects: high-energy radiation and low-energy radiation. The term "high-energy radiation," or in scientific circles, "ionizing radiation," is used when the radiation energy level is high enough to tear the tiny charged particles known as electrons away from the atoms which make up all life forms. Once the electron is removed from the atom, the electrical neutrality of the atom is lost. The departing electron takes its negative charge with it and the atom is left with a positive charge making it an ion, hence the term "ionizing radiation." Because ionization has a critical impact upon our health, much of the discussion that follows will focus on the high-energy forms of radiation.

High-Energy (Ionizing) Radiation

Cancer

Except for massive doses of radiation which can cause sudden death, cancer and genetic mutation are the prime concerns associated with high-energy radiation. There is now no doubt among scientists that high-energy (ionizing) radiation can cause cancer. There is a lot of debate, however, about precisely how radiation causes cancer, and about the likelihood of cancer resulting from a given radiation exposure.

Researchers have constructed a variety of competing models to try to explain the relationship between levels of radiation and the incidence of cancer. These models and the reasons for the large degree of uncertainty are discussed in chapter 6. Because of the uncertainty and lack of unanimity among researchers, the prudent approach is generally to assume that the risk of cancer is directly proportional to the amount of radiation exposure.

This means that there is *no safe dose* for radiation. Every small dose has an associated risk and, except for massive doses, it appears not to matter if the exposure is spread out over a period of time. It also means that there is no threshold dose below which we need not worry. The fact that a dose of radiation is small, or even less than the dose we all get each year from natural sources, does not make it harmless.

The other key point is that spreading the dose around, for example, by releasing radioactive gases into the environment, does not reduce the amount of cancer caused. All it does is spread the risk around— any one individual will have a lower risk of cancer but the total number of cancers caused will be the same. An example will help illustrate the point: If a particular job at a nuclear plant involves exposing a worker to enough radiation to give a one-in-ten chance of cancer, sharing that job (and the radiation exposure) among ten workers will reduce each worker's risk to one-in-one hundred but the total risk will be the same—a one-in-ten chance that one of the workers will get cancer.

While there is some debate about the relationship between the age at which a person is exposed and the risk of cancer, there appears to be a greater susceptibility among the young. Unborn children in the womb may be the most susceptible, especially in the early months of pregnancy.

Due to the long latency period for cancer (the period of time between the exposure to radiation and the recognition of cancer), cancers are seldom visible before at least a decade has passed, and often take several decades to show up. This means that the exposure of older people to radiation may be less serious because the cancer may never surface in their lifetime. It also means that when cancer does appear, it is difficult to know what in particular caused the cancer.

Genetic Effects

High-energy radiation does not just create a risk for the person exposed to it. By damaging our genes it can indirectly affect our children, grandchildren and even their children. The exact level of risk is unknown but the seriousness of the effects, including possible deformities, retardation and illnesses ranging from schizophrenia to cancer, means that such risks are not to be treated lightly. But since the risk of genetic mutation is imposed on future generations it leads to a different set of concerns.

In theory, if the risk of cancer due to a given dose of radiation was known and fully explained to a worker or consumer or medical patient and the exposure was voluntary, there would be little reason for concern on the part of anyone other than the individual affected. The person involved could weigh the risk and the benefits and come to a sound conclusion based on his or her own values and priorities. But when a person of reproductive age is exposed to the risk of genetic mutation because of radiation, should he or she make a decision affecting future generations?

Similarly, when we decide to pollute the environment with radioactive substances from nuclear power production, it is conceivable that we as a society may believe that the benefits of nuclear power outweigh the cost in cancer for the present generation. But do we have a right to hurt future generations by leaving around long-living radioactive waste which will cause cancer among our descendants?

Questions such as these do not lend themselves to simple answers. We may all agree that given the benefits of radioactive cobalt therapy for cancer victims it is worth the risk created when we manufacture the cobalt. But to know whether that conclusion is a good one, we must understand more about the risks, benefits and present degree of uncertainty about the effects of radiation.

High-Energy Radiation—Estimates of Risk

High-energy radiation doses are usually expressed in units called "rems." The number of rems is the measure of the biological harm caused by radiation to each gram of body tissue exposed. To know the risk of cancer or genetic mutation we must know what part of the body is exposed and how many rems each exposed area received.

One rem to the whole body means that each and every gram of tissue in the body receives an average dose of one rem. Therefore, one rem to the whole body is far more serious than one rem to the fingernail. Also, one rem to a sensitive organ such as the lungs is of more concern than one rem to a part of the body which is less likely to become cancerous.

Of course, a dose to a pregnant woman endangers both mother and child, and exposure of the sex organs to high-energy radiation can cause the risk of genetically inherited disorders in future generations.

The terms "millirem" (one thousandth of a rem), "rad" (similar to rem for most forms of high-energy radiation) and "millirad" (one thousandth of a rad) are most commonly used; *see* Glossary for System Internationale equivalents and definitions.

A number of official and quasi-official scientific bodies and researchers have estimated the risk of cancer and genetic injury due to each rem of high-energy radiation. These estimates are expressed as the number of cases of cancer or genetic damage caused by exposing one million people to one rem of radiation to their whole bodies. (*See* Table 1.)

Since the total risk is the same if one person or one million people share the same total dose, these risk estimates can be used to calculate your risk due to a given exposure. For example, taking Gofman's number of 3771 fatal cancers for one million people each receiving 1 rem, we can also say that one person receiving 1 rem will get one-millionth of 3771 cancers or we might more properly say that that person has 3771 chances in a million of getting fatal cancer from 1 rem of radiation (roughly 1 in 250). And since each bit of exposure has a corresponding risk, ten rems would lead to one chance in twenty-five.

An important point to keep in mind when comparing radiation doses to these risk estimates is that these are risks for *whole-body exposure*. Partial-body exposures (from localized X rays, for example) create less and more identifiable risk (exposure of the lungs leads to a risk of lung cancer, where exposure of the whole body creates a risk for many types of cancer).

In order to make it easy to compare the relative risks of different partial-body exposures, the unit "Effective Dose Equivalent" (EDE) or "Whole-Body Dose Equivalent" is used. This unit accounts for the lower risks of partial-body exposure and for the particular radiation sensitivity of the parts of the body exposed. In other words, the Whole-Body Dose Equivalent is the dose to the whole body that would give the same risk as the dose actually received by a part of the body. Thus a dose of 10 rems to the pancreas may be expressed as a Whole-Body Dose Equivalent of 1 rem (giving a

TABLE I

High-Energy Radiation Risk Estimates

(for exposure of 1,000,000 persons to one rem to the whole body)

Committee or Researcher	Fatal Cancers	Nonfatal Cancers	Genetic Damage	Total
Intl. Comm. on Radiological Protection (ICRP), 1977	125	125	40–80	290–330
U.N. Scientific Comm. on the Effects of Atomic Radiations (UNSCEAR), 1980	100–200	111–332	20–150	231–682
U.S. N.A.S. Comm. on the Biological Effects of Ionizing Radiation (BEIR), 1980*	28–501	4–770	24–440	56–1710
J. Gofman, 1981	3771	n/a	191—>20,000	3962—>23,962
R. Bertell, 1982	384–1450	121–494	500–10,600	1005–12,544

* Relative risk method, range includes different models considered.

1-in-250 chance of fatal cancer according to Gofman). Researchers convert a partial-body exposure to its EDE based on research into the incidence of cancer for the particular body part exposed *versus* the total incidence of all types of cancer in the population. The relative incidence of localized cancers and the corresponding sensitivity of various parts of the body are discussed in greater detail in Chapter 6.

Finally because we believe in a philosophy of prudence when there is great uncertainty about the effects of certain substances on the public's health, and because we are impressed with Dr. John Gofman's methodological rigor and clarity and his independence from vested interests, this book relies on Gofman's estimates for cancer risks due to high-energy radiation.

Low-Energy (Non-Ionizing) Radiation

We noted earlier that the biological harm caused by radiation tends to correspond to its energy level—in general, the higher the energy (or frequency), the more harmful the radiation. We also noted that the spectrum ranging from extremely low energies and frequencies to very high ones breaks into two very different sections depending on whether the energy level of a particular form of radiation is high enough to seriously disrupt the structure of a cell by ionization. Generally speaking, then, low-energy radiation is less worrisome than high-energy radiation.

But before we complacently settle back in front of our televisions, we should look into the subject a little more deeply. It was not so long ago that our ignorance of the health effects of high-energy radiation allowed us to indulge in what now seem to be ludicrous uses of radiation. For example, X-ray fluoroscopes were used to check children's shoe sizes in department stores, no doubt leaving a small but very real legacy of cancer.

Since then many scientists who became outcasts in the scientific community because of their warnings about the risks of relatively small exposures to high-energy radiation have been slowly reinstated to respectability. Exposure limits have been repeatedly tightened, and unnecessary exposures greatly reduced, although many consumers, workers and patients were still exposed to large

TABLE 2

Annual Dose Rates from Common Significant Sources of High-Energy (Ionizing) Radiation Exposure

Source	Exposed Group	Body Portion Exposed	Average Dose Rate (millirems/yr)
NATURAL BACKGROUND			
Cosmic radiation	Total population	Whole body	28
Terrestrial radiation	Total population	Whole body	26
Internal Sources	Total population	Gonads	28
		Bone marrow	24
MEDICAL X RAYS			
Medical diagnosis	Adult patients	Bone marrow	103
Medical personnel	Occupational	Whole body	300–350*
Dental diagnosis	Adult patients	Bone marrow	3
Dental personnel	Occupational	Whole body	50–125*
RADIOPHARMACEUTICALS			
Medical diagnosis	Patients	Bone marrow	300
Medical personnel	Occupational	Whole body	260–350
ATMOSPHERIC WEAPONS TESTS	Total population	Whole body	4–5
NUCLEAR INDUSTRY			
Commercial nuclear power plants (effluent releases)	Population within 10 miles	Whole body	<< 10
Commercial nuclear power plants (occupational)	Workers	Whole body	400†

Source	Exposed Group	Body Portion Exposed	Average Dose Rate (millirems/yr)
Industrial radiography (occupational)	Workers	Whole body	320
Fuel processing and fabrication (occupational)	Workers	Whole body	160
Handling byproduct materials (occupational)	Workers	Whole body	350
Federal contractors (occupational)	Workers	Whole body	~ 250
U.S. Naval nuclear propulsion program (occupational)	Workers	Whole body	220
RESEARCH ACTIVITIES			
Particle accelerators (occupational)	Workers	Whole body	Unknown
X ray diffraction units (occupational)	Workers	Extremities and whole body	Unknown
Electron microscopes (occupational)	Workers	Whole body	50–200
Neutron generators (occupational)	Workers	Whole body	Unknown
CONSUMER PRODUCTS			
Building materials	Population in brick and masonry buildings	Whole body	7
Television receivers	Viewing populations	Gonads	0.2–1.5

Continued over

Source	Exposed Group	Body Portion Exposed	Average Dose Rate (millirems/yr)
MISCELLANEOUS			
Airline travel (cosmic radiation)	Passengers	Whole body	3
	Crew members and flight attendants	Whole body	160
Airline transport of radioactive materials	Passengers	Whole body	~ 0.3
	Crew members and flight attendants	Whole body	~ 3

*Based on personnel dosimeter readings; because of relatively low energy of medical X rays, actual whole-body doses are probably less.
†Average dose rate to the approximately 40,000 U.S. workers who received measurable exposures was 600–800 mrems/yr.
Source: BEIR (1980). See Appendix C

doses even when evidence of the risks of high-energy radiation was already surfacing.

Surely today we can do better than to follow this previous pattern with exposure to low-energy radiation.

Today our knowledge about the effects of low-energy forms of radiation is at the same level as the knowledge of high-energy forms was forty years ago. We do know that large acute doses hurt, just as acute doses of high-energy ionizing radiation do. But very little investigation of the less obvious effects of prolonged low-level exposure has been done.

At the dose levels we typically experience in the environment (for example from telecommunications) or in the home (from micro-wave ovens) or in the workplace (from video display terminals), low-energy forms of radiation do not result in immediately identifiable damage to the cells in our bodies. The primary concern with low doses of low-energy radiation is the possibility of a subtle effect on our nervous systems and chemical "communications systems" within our bodies that regulate virtually every bodily function. With prolonged exposures some delicate systems may be totally disrupted, causing them to fail or increasing the exposed persons' susceptibility to disease. Such a disruption of nervous and chemical feedback systems is particularly dangerous for a developing fetus.

While it appears that low-energy radiation may be linked to cancer or genetic defects, we cannot rely on our knowledge of high-energy radiation effects to estimate risk, because the nature of the interaction between low-energy radiation and our bodies is quite different. Low-energy radiation does not ionize the atoms in our cells. The units of measurement for low-energy radiation must therefore be different since terms such as "rem" consider the degree of ionization. A dose of low-energy radiation is, therefore, usually expressed as a "field strength" experienced over a period of time and, unlike the rem, is not a direct measure of biological disruption.

Finally, the pattern of low-energy radiation — whether pulsed or continuous — appears to be an important factor as pulsed forms appear to be more likely to do damage (*see* Low-Energy Radiation, Form *versus* Amount in Chapter 7).

TABLE 3

Some Uses of Low-Energy Radiation

Typical Frequency	Use
300,000 MHz	Microwave relay. Short-range military communications
30,000 MHz	Commercial satellites. Direct-broadcast TV satellites. Microwave relay. Military communications. Air navigation. Radar
3000 MHz	UHF television. Police and taxi radio. Microwave ovens. Medical diathermy. Radar. Weather satellites
300 MHz	FM radio. VHF television. Police and taxi radio. Air navigation. Military satellites
30 MHz	International shortwave. Air and marine communications. Long-range military communications. Ham radio. CB
3 MHz	AM radio. Air and marine communications. Ham radio. SOS signals
0.3 MHz	Air and marine navigation
0.03 MHz	Time signals. Military communications. Weapons and theft-detection scanners
0.003 MHz	Electric power (AC). Military communications. Electric transportation systems
0 MHz	Electric power (DC). Batteries. Bone stimulation

For a detailed discussion of the health impacts of the various low-energy radiations see: Microwaves and Radiowaves, and High-Voltage and House-Current ELF Radiation.

The Concept of Risk

What It Means

Perhaps the most important question to answer about a given form and dose of radiation is whether or not it hurts you. But the answer to this question, even if radiation and radiation-induced disease were completely understood, can rarely be more definite than "probably." Except for very high doses of radiation we can only speak of the *chances* of being hurt.

Take the analogy of a blind pedestrian crossing a road: We would all agree the pedestrian is at some risk. We might even put a number on the risk, say a one-in-one-hundred chance of his being hit. Does this mean the pedestrian will necessarily be hit? Clearly not. Nor does this mean that if the pedestrian crosses the street one hundred times there must be a collision or if one hundred blind pedestrians cross the street once each, that one of them must be run over or that two of them would not be. But we can say that given a very high number of such crossings, on the average there will be one accident for each one hundred crossings.

A Matter of Apples and Oranges

When asked how concerned a resident should be about living near a nuclear power plant, the typical response of a plant spokesperson is, "Let's put the risks into perspective." He then goes on to compare the amount of radiation leakage to a fraction of the amount received during a chest X ray and suggests that the chances of an accident are about the same as the odds of being hit by a meteorite.

While the spokesperson's statements may be accurate, they are also dangerously misleading if they lead the residents around the power plant to conclude that there's nothing to worry about.

A closer examination of all the factors to be considered when weighing risk will show why.

Probability and Consequences. Risk is clearly a matter involving two major elements: *probability* (that is, how likely this problem is to occur) and *consequences* (how bad it would be if it did). Most scientists and engineers are taught to simply multiply together the best estimates of probability and consequences, and to treat that mathematical product as if it were the only thing that mattered.

Engineers typically think we should always avoid a large product (which they call a large risk) and prefer a small product (which they call a small risk), but the normal, prudent thing to do is to avoid, wherever possible, risking unacceptable consequences, even if they are unlikely. To our ancestors, that meant feeding the new mushroom to the dog before eating it. To us, it means protesting when somebody wants to put a chemical factory, or a waste dump, or a nuclear plant in our neighborhood.

Thus, most people would gladly suffer a sure nuisance in order to avoid an unlikely disaster, in violation of the engineers' simple formula. This happens, for example, every time we buy fire insurance, and pay money year after year in order to avoid the unlikely possibility of being bankrupted by a fire.

In addition, most people are naturally wary of pseudo-scientific assurances that all will be well. It is easier to believe that certain technologies (like pesticide factories or nuclear reactors) have the potential for a disaster if things go badly, than it is to believe that nothing will go wrong. And judging by experience, the public is right to be skeptical of "experts" who try to allay their concerns with promises that possible consequences will never (or virtually never) happen.

Voluntary *versus* Involuntary Risk. The imposition of risk upon another person adds both ethical and psychological elements to the equation. Who has the right to create risk for another person? There seems to be little justification for taking the decision to accept or reject risk away from the person who must live with it, but as a society we do so regularly. When a person is not told about the hazard associated with a radioactive consumer product or a medical technique involving radiation, the right to informed decision-making is nonexistent. When a utility mines uranium and uses it in a nuclear power station, the risk is borne by all who live nearby—whether they like it or not. And when these risks place the health of future generations in peril, those particular victims cannot be polled for their views.

Psychologists tell us that voluntary and involuntary risks are perceived differently with a voluntarily assumed risk being much more acceptable to people. This consideration accounts for some of the differences between the risk of radiation from a nuclear plant, which is for most people an involuntary risk, and the risks of the same radiation dose from an X ray, which, if the patient is well informed, is a voluntary risk. The availability of an acceptable alternative and the benefits that flow from taking the risk, also affect the choice.

Acceptability, Alternatives and Benefits. We may be willing to live with the risk associated with a technology such as the medical X ray so long as no other approach is available, but once an equally effective technology with less risk is found, the acceptability of the X ray disappears. The nuclear industry spokesperson who fails to mention low-risk alternatives to nuclear power, such as energy conservation, ignores this important distinction.

Benefits. When nuclear plants are equated with medical X rays, there is never any mention made of the benefits of each. X rays often save lives; nuclear power plants never do. Nor do ceramic dinner plates with radioactive glazes. The nature of the benefits associated with a particular level of risk varies dramatically from one example to the next. In the sections of this book dealing with specific sources of radioactivity these distinctions have been made and should be borne in mind by the reader.

Additive Risk. Finally, when evaluating the risk of radiation-induced disease it should be of little comfort that any particular dose is less than, or roughly the same as, the dose we all receive from natural background radiation, because it is in addition to that radiation, not instead of it. Getting hit once with a baseball bat does not make a second beating less offensive. Consequences and the risk of consequences are additive—every little bit hurts.

Common Forms of Radiation—Their Properties and Associated Problems

Low-Energy Radiation (Non-Ionizing)

ELF (extremely low-frequency) radiation
- a form of non-ionizing radiation
- very low energy
- health effects not understood
- possible carcinogen
- possible behavioral effects
- sources: electric power lines, submarine communications, video display terminals.

RADIO WAVES (RF or Radio Frequency)
- non-ionizing
- very low energy
- health effects not understood, suspected of disrupting physiological processes including cardiovascular system
- sources diverse, including radio and television transmission.

MICROWAVES
- non-ionizing
- low energy
- at low levels suspected of disrupting physiological process
- at high levels can raise body temperatures and cause cataracts
- sources: microwave ovens, telecommunication towers, radar.

INFRARED
- invisible, non-ionizing
- low energy
- at high levels can raise body temperatures
- sources: all warm objects, the sun, heat lamps.

VISIBLE LIGHT
- visible, non-ionizing
- low energy
- at high levels can raise body temperatures and cause eye damage.

ULTRAVIOLET (UV)

- invisible, can cause some ionization
- low energy
- can cause sunburn, eye damage, skin cancer
- suspected of causing behavioral effects including hyperactivity in children
- sources: sunlight, fluorescent lights, tanning lamps.

High-Energy (Ionizing) Radiation

ALPHA PARTICLES

- ionizing particle emitted by some radioactive substances (e.g., uranium and radium)
- positively charged, made of 2 protons and 2 neutrons
- medium-to-high energy
- large size of alpha particle limits its ability to penetrate body tissue
- particularly dangerous when emitted within the body, from swallowed or inhaled radioactive materials
- can cause cancer, leukemia, birth defects (if child exposed in uterus) and genetic damage
- due to their weight and charge, alpha particles within the body are roughly 10 times more ionizing—and more damaging—than other forms of radiation with a similar energy level
- sources diverse, including: uranium mining, radon gas, mantle lamps, cigarette smoke.

BETA PARTICLES

- ionizing
- fast-moving particles emitted from some radioactive substances (e.g., tritium)
- negatively charged, made of high-speed electrons
- medium-to-high-energy, can penetrate 1—2 cm into water or flesh
- can cause cancer, leukemia, birth defects (if child exposed in uterus) and genetic damage
- sources diverse, including: nuclear power plants, eyeglasses, dental porcelains.

X RAYS
- ionizing
- highly penetrating
- medium-to-very-high energy
- can cause cancer, leukemia, birth defects (if child exposed in uterus) and genetic damage
- man-made, used in medical sciences and industry, emitted by some TVs and VDTs.

GAMMA RAYS
- similar to X rays but emitted by radioactive materials and usually of higher energy
- sources diverse, including: nuclear medicine, nuclear power plants, building materials.

NEUTRONS
- ionizing particles
- highly penetrating
- high energy
- can cause cancer, leukemia, birth defects (if child exposed in uterus) and genetic damage
- can interact with other material inducing radioactivity
- sources: cosmic radiation and nuclear reactors.

2

RADIATION: WHERE IS IT?
WHAT CAN YOU DO
ABOUT IT?

CHAPTER TWO

Radiation

in the Household

For many of us the word "radiation" brings to mind an image of workers clad in protective space suits. This image places the mysterious rays in the realm of high tech—isolated in the laboratory or buried behind the concrete bunkers of a nuclear reactor core.

But we need not venture into such alien environments to be exposed to the hazards of radiation—the kitchen and den will do nicely.

We are all exposed to household sources of radiation. Most emit relatively small amounts, but, nevertheless, they create a health hazard that should be recognized and weighed against the benefits of the particular appliance.

An examination of common sources of radiation, their degree of radioactivity and of alternative ways of meeting the same needs will allow you to minimize the risk to yourself, to your family and to others sharing the environment.

Keep in mind that even small dose rates can damage your health, especially when the exposure is prolonged as it often is with household sources.

Ceramic Tableware and Glassware

One of the best illustrations of our ignorance of the health effects of radiation is the use of uranium compounds in ceramic glazes for tableware and in the coloring agents for decorative glassware. Recently some brands of tableware have been found to have glazes containing as much as 20 percent uranium oxides by weight. The uranium gives off high-energy gamma and beta radiation. Exposure is internal as well as external since many acidic foods leach the compounds out of the glaze and are subsequently swallowed.

Typical external exposures on contact are in the 5 to 10 millirem per hour range. At this rate, using the plates for three meals a day would give a dose in excess of the regulatory limit for exposure to non-medical radiation sources for the general public and may cause skin cancer. Swallowing the dissolved oxides can be hazardous, particularly to the kidneys where the metal concentrates. It is facts like these which have caused the U.S. Bureau of Radiological Health to recommend that use of uranium-based glazes be banned.

Unfortunately a great many of these dishes were manufactured, most prior to the 1940s, though some as recently as 1978. Many of the pieces are collected as examples of art deco, the most common being "Fiesta red" which was the most popular color produced by Homer Laughlin China Co. of West Virginia, the supplier of Fiestaware to both the U.S. and Canada from 1935 to 1971.

The glazes are characterized by a shiny reflective surface, often orangy-red, beige or yellow. Brands identified as being possible suspects include:

Fiesta
Red Wing
Yellow and pink Franciscan
Early California
Green Catalina
Orange Medicine Hat (Matina Ware, Indian Head Marking)
Medalta (Orange Swirl Pattern)
Edwin M. Knowles China Co.
Harlequin, Vistosa, Caliente, Riveria, Stangl and Poppytrail
   by Metlox

If in doubt, contact your local health authority to arrange for a test of your dishes. Clearly these dishes should not be eaten from and, for most, the decorative value is not enough to outweigh the risk from radiation exposure. If you have eaten regularly from such dishes for many years ask your physician to be alert to the possibility of skin cancer or kidney disease. Decorative glassware containing radioactive agents is less common and not usually used for food service, and therefore of much less concern.

Cigarettes and Tobacco

For years scientists have been aware of the dangers of cigarette smoke, especially with regard to lung cancer and atherosclerosis (the buildup of fatty deposits on artery walls). Other risks include cancer in organs distant from the lungs, such as the bladder and the pancreas.

While most research has centered on isolating chemicals in tobacco smoke which are carcinogenic, recently a new theory has emerged to help explain the diverse health effects of cigarette smoking. It focuses on the fact that tobacco smoke contains radioactive particles.

The radioactivity of tobacco smoke is not in itself a new discovery. As early as 1964, Harvard University researchers noted high concentrations of lead 210 and polonium 210 (radioisotopes which result from decay of naturally occurring radon) on the tiny hairs called trichomes that abound on tobacco leaves. (The substances may come from phosphate-based fertilizers used by tobacco growers.)

Because these radioisotopes are alpha emitters they were always of some concern, but their low concentrations in the smoke did not explain the high incidence among smokers of diseases throughout the body. Most smoke particles which are inhaled dissolve in lung fluid and then rapidly clear, allowing little time for irradiation of lung tissue by the radioactive particles.

Recent research has shown, however, that Pb 210 (a radioisotope of lead) is found in smoke particles that do not dissolve in the lung. The insoluble particles remain in the lung for roughly two years and also move through the body to the cardiovascular system, lymph nodes, liver, spleen, bone marrow and other sites where they can remain for many years. The lead 210 decays over time to bismuth 210 and polonium 210 which emit alpha radiation.

Researchers have found high levels of lead 210 and polonium 210 in the tumors of smokers. High levels have also been noted in the fatty arterial deposits characteristic of atherosclerosis. Although the carcinogenic quality of alpha radiation is clear, its role in the development of atherosclerosis as yet is not understood.

The effects of cigarettes upon health should be no surprise to smokers, and the fact that in addition to the chemical carcinogens radioactivity may be an "active ingredient" in smoke may be of little concern to the addicted smoker. The recent findings do, however, support the view that all low-level internal exposures to alpha radiation can be a serious health concern.

Cloisonné Jewelry

Like uranium-bearing glazes on tableware and glassware (*see* Ceramic Tableware and Glassware), some jewelry contains radioactive uranium or thorium and can give the user a substantial dose of radiation. Although the enameled jewelry known as cloisonne, may be more harmful than the tableware, it has received less attention, presumably because it is less common.

The Nuclear Regulatory Commission has found that approximately 10 per cent of cloisonne jewelry emits 3 to 7 millirems of high-energy radiation per hour measured at its surface.

If the jewelry has no backing, as is sometimes the case with pendants, the wearer could receive a dose to the skin of 2000 to 4000 millirems per year if the jewelry is worn on an average of 10 hours per week.

If a backing is used on the jewelry, the dose is substantially reduced to the 25 millirems per year level. Nevertheless, the risk of skin cancer from wearing the jewelry cannot be considered acceptable given the minimal benefit of the product.

In the U.S., the Nuclear Regulatory Commission suspended the importation of uranium-bearing cloisonne as of July 25, 1983, but this will not stop imports of thorium-bearing cloisonne which is also dangerous or the sale or use of radioactive jewelry already in circulation.

Owners of cloisonne jewelry should avoid wearing it unless they are satisfied that the particular article is not radioactive.

Eyeglasses

Eyeglass lenses contain rare earths added to enhance optical quality. These rare earths and the silica which makes up the bulk of the glass often contain radioactive thorium or uranium. These elements decay into a number of radioactive substances and in the process emit alpha, beta and gamma radiation.

In the U.S., limits have been placed on the uranium and thorium content of opthamalic glass. This limit results in a maximum dose of approximately 4 rems per year to the outer layers of eye tissue due to alpha irradiation. The beta dose is far lower, as is the gamma, although it penetrates farther into the body than alpha radiation. This dose rate is based on use of the glasses for 16 hours each day.

The cancer risk from this source of radiation is probably slight but there is evidence that high doses of radiation can cause cataracts. Eyeglass wearers should consider using plastic lenses which, while less scratch resistant and slightly more expensive, are lighter and nonradioactive.

Dental Porcelains

Until the mid-1970s, porcelains used to make dentures (false teeth) and found in bridges and crowns usually contained uranium and cerium which give a fluorescent quality similar to that of natural teeth.

The porcelains usually contain between 100 and 500 parts per million of uranium, and give off alpha, beta and gamma high-energy radiation. The gamma dose is extremely low and the alpha particles, while given off in significant quantities, do not usually penetrate the skin deeply enough to result in damage to sensitive tissue. The beta particles, however, do penetrate to the basal cell layer of the skin, a type of skin cell which can become cancerous.

The Nuclear Regulatory Commission estimates typical doses to the mouth area of 1.0 rem per year for a person wearing a porcelain tooth. Using Gofman's estimate of cancer dose we would expect roughly one cancer death for each year of use among each one hundred thousand people. The cancer risk has prompted the British government to recommend the discontinuation of the use of radioactive fluorescers in the manufacture of dental porcelains. Alternatives include nonfluorescent porcelain and plastic compounds. While the lack of the fluorescence may make the denture appear slightly unnatural in color, especially under artificial light, for most people the cosmetic value does not warrant the health risk.

Fluorescent Lighting

It's not uncommon to hear people blame their irritability or eye strain on fluorescent lighting.

The slight flicker of the tubes may be one cause for complaint, but recent studies have found that the light radiation itself may damage our health.

Fluorescent lighting found in most homes, schools and workplaces does not have the same spectral makeup as sunlight. While sunlight itself can cause or contribute to skin cancer, the differing pattern of fluorescent light may be an even greater worry. Compared to the sun, a typical cool white tube emits an overabundance of long-wavelength ultraviolet light and a relative lack of a range of light frequencies including shorter wavelength ultraviolet.

The ultraviolet (U.V.) light (also found in sunlamps) is considered beneficial only in very small doses as an aid in vitamin production. In larger doses it damages the skin and is associated with skin cancer. A recent Australian study found that women who worked under fluorescent lights were twice as likely to develop malignant melanomas (a potentially fatal form of skin cancer) than women who had little exposure to fluorescent light. Men were found to be at even greater risk. One interesting finding is a prevalence of these melanomas on the shoulders and trunk of the body. It is believed the ultraviolet light penetrates thin and loosely woven garments, exposing skin surfaces which are less likely to be tanned. The slight tan most people have on their hands and face is believed to be protective.

The present incidence of skin cancer in the U.S. is roughly three to four hundred thousand new cases per year, with approximately five thousand deaths per year due to the malignant melanoma variety.

The high concentration of some ultraviolet wavelengths may not be the only problem with fluorescent light.

One respected photobiologist, Dr. John Nash Ott, who has studied the effects of differing light spectra on animals and plants for over fifty years, believes hyperactivity in young children may be one effect of fluorescent lighting.

Other studies have tended to confirm a relationship between hyperactivity and fluorescent lights but there is some debate as to whether the effect is due to the light radiation or very low levels of X-ray radiation leaking from the ends of the tubes.

A study carried out at the Massachusetts General Hospital found that the calcium absorption of men living under fluorescent light fell by 25 percent during the winter.

In studies on animals it has been found that fluorescent light contributes to tooth decay. Dr. Ott believes many of these effects may be due to the disparity between natural and fluorescent lighting. He has demonstrated effects upon the pineal and pituitary glands which regulate numerous functions in our bodies, due to exposure of the eyes to different frequencies of light. Dr. Ott speculates that it is the absence of some wavelengths in fluorescent light that may cause the problem.

Many of these potential problems can be reduced or avoided. Incandescent lamps, while far less energy efficient, do mimic the natural spectrum much better than the fluorescent variety.

Broad-spectrum fluorescent lamps are also better at providing missing wavelengths but do not reduce all wavelengths of ultraviolet light exposure. Diffusing covers over fluorescent tubes are a mixed blessing. They limit both longwave and shortwave ultraviolet light by as much as a factor of 10. This reduction is recommended to reduce the risk of cancer but if Dr. Ott is correct, the loss of some of the wavelengths may lead to more subtle nervous-system problems. Shielding the ends of the tubes with metal casings will reduce X-ray leakage, and many fluorescent fixtures have such a metal housing.

Probably the best way to deal with the problem, and still maintain energy savings, is to place incandescent lights near work areas rather than bathing the entire room with artificial light.

Irradiated Food

The press has paid a good deal of attention to recent proposals calling for widespread irradiation of food as a simple and cheap sterilization and preservation technique. The media is no doubt responding to the uneasiness most people feel when they first hear of food and irradiation in the same breath. Yet our fears of "hot" food are unfounded—the process does not make food radioactive.

There are, however, other legitimate concerns with the process including its effects on the quality of food and the dangers of the irradiation process itself both for the workers involved and for the environment in general.

The process usually involves exposing foods to very high dose levels of gamma radiation, up to one thousand times the dose that would kill a person. Cobalt 60 is the most commonly used source of the radiation. It is produced by exposing nonradioactive cobalt to the neutrons in the core of a nuclear reactor. Accordingly food irradiation entails all the hazards associated with the widespread production and handling of cobalt 60—hazards such as the inevitable leaks due to accidents as cobalt 60 is transported to hundreds of food-processing plants as well as the dangers of improper waste disposal.

While food irradiation is presently quite restricted in both Canada and the U.S., broader licensing is imminent and the technique is already widely used in the Netherlands, Israel and South Africa. In North America, the irradiation of potatoes to inhibit sprouting has been permitted for many years but has not proven to be useful due to problems of contamination after treatment.

Vitamins A, B1, B6, B12, C and E are sensitive to irradiation although alternative preservation techniques such as canning and precooking also destroy them to some degree. Very high doses of radiation can reduce the nutritional value and in some cases even produce toxic substances—a factor one hopes regulators will consider in determining permissible limits.

Animals fed irradiated food have been found to have some cells with abnormal chromosome counts, a condition known as polyploidy. Questions about the likelihood and seriousness of polyploidy in humans have yet to be laid to rest.

Offsetting these concerns are the benefits of reduced use of the currently available alternatives including such carcinogenic chemicals as EDB (ethylene dibromide). This benefit is limited by the fact that irradiation provides no protection from recontamination after treatment whereas chemical additives do. Irradiation may also hide viral infestations because it kills bacteria that often accompany viruses and which would ordinarily give off telltale odors.

Unfortunately, as in the case of chemical additives, there is very little understanding of the subtle long-term effects on human health due to food irradiation, and we may all become experimental guinea pigs. Foods which have been irradiated will likely bear a euphemistic notice such as "TREATED WITH IONIZING ENERGY" if current proposals become law. This will allow you to identify these products if you wish to avoid them until more is known about their effects.

Mantle Lamps

Most campers and cabin owners are familiar with the bright white light of a gas, propane or kerosene mantle lamp. The suspension of a delicate mesh, or mantle, over the flame boosts the light output of these lanterns by a factor of 4 compared to the wick variety, but it also adds a potentially serious radiation source to the otherwise comforting country setting.

The mantles contain between 250 and 400 milligrams of thorium which when heated glows brightly, giving off the familiar white light.

Thorium is an alpha-emitting radioactive substance which decays over time into a variety of radioactive daughters. A daughter is the element a radioactive substance becomes when it decays. Daughters can, in turn, decay if they are radioactive themselves. The first daughter produced in this decay process is radium 228—a beta emitter. Radium 228, in turn, decays to alpha-emitting radioactive daughters and radon 220.

The alpha particles are usually stopped by the glass shade of the lamp once the mantle is installed but inhaling or swallowing part of the mantle or its ash can lead to potentially damaging internal exposures. The beta radiation while not highly penetrating is of some concern to persons handling the mantles in order to install or remove them. Mantles should not be carried in shirt or pants pockets for great lengths of time, and should be kept out of the reach of children.

Much of the radium which accumulates from thorium decay is released to the air in the first 15 to 20 minutes of each burning. To reduce exposure the lamps should be kept outside or in a well-ventilated area during this period, especially during the initial "preburning" of a new mantle. This is especially important for tent and trailer campers.

Changing the mantles is a particular problem since the chance of swallowing or inhaling the ash is relatively higher. Care should be taken not to inhale the ash or spill it near food. Ashes should not be disposed of in a compost heap or tilled into garden soil.

Typical annual exposures are as follows:

TABLE 4

	Millirems/year*
Direct Exposure	0.0012–0.024
Mantle Replacement	0.25 –1.3
Radon 220 Emanation	0.052 –1.0
Total	0.3032–2.324

Ranges based on ten-to-two-hundred days per year exposure due to typical use of one lamp.

Such exposures would have corresponding average fatal cancer risks of between 1 and 9 chances per million for each year of exposure, not a serious concern when compared to the situation of a child swallowing a used mantle. Such a child would be exposed to roughly 16,000 millirems over the following 50-year period with an approximate 1 percent risk of fatal cancer.

Cancer risk is highest for bones and lungs, and genetic damage is a concern, particularly if the mantle is swallowed.

The mantles also contain berylium to give them more strength once preburned. Half of this berylium is vaporized in the initial 15-minute burn. While not radioactive, berylium is a cause of lung disease and is another reason to conduct this procedure out of doors.

Given the availability of alternatives to mantle lamps such as wick lamps and flashlights the risk is relatively high.

Microwave Ovens

Microwave ovens use high levels of microwave radiation to add energy to food and thereby cook it. The food does not become radioactive. Metal casings and screens built into the ovens should stop the radiation but the ovens can sometimes leak, especially at the door seal.

U.S. regulations permit only 1 milliwatt per square centimeter of leakage at the time of manufacturing but up to 5 milliwatts per square centimeter after sale. The "after sale" limit is five times the recommended maximum dose rate for chronic public exposure in Canada.

Keeping the door seal clean will substantially reduce the amount of leakage. The ovens should never be operated while empty as this will increase leakage and children should be discouraged from peering through the window to watch what is cooking.

Virtually every restaurant or fast-food establishment has a microwave oven to cook or reheat food. Because of possible leaks restaurateurs would be wise to locate them in a place where the time workers spend standing or sitting very close to the oven is reduced.

A variety of inexpensive microwave leakage testers are available from electronics distributors which will alert users to substantial leaks. The devices were tested by the Consumers Association of Canada (November 1981) and two, priced at less than twenty dollars, were found to measure radiation at levels down to 1 milliwatt per square centimeter:

MICRO-CHECK
(developed by the National Research Council of Canada)
Bow Rae Sales & Services Ltd.,
Westwind Industrial Park,
Box 9139, Station E,
Edmonton, Alberta,
T5P 4K2.

MICRO-PROBE
(Australian made), distributed by Westclox Canada,
P.O. Box 239,
Peterborough, Ontario,
K9J 6Z1.

Radioactive Food

In the 1950s and 1960s, many mothers resorted to using powdered milk for their families during the periods following the atmospheric testing of nuclear bombs. Today these devices are usually exploded underground and the radioactive fallout is more contained.

However, the fallout from the earlier tests is still with us, though in gradually decreasing quantities. Little can be done to avoid such long-lived carcinogenic radioisotopes as strontium 90 which concentrates in our bones and cesium 137 which concentrates in soft tissues such as the muscles and reproductive organs with a corresponding cancer risk.

However, there is presently another potential source of radioactive food—food grown in the shadow of nuclear power plants. The most serious example is the cultivation of tomatoes in greenhouses heated by steam from nuclear power plants. The practice has received backing from at least one utility in Canada, Ontario Hydro, which is anxious to find a profitable market for the enormous supply of waste heat produced by its many nuclear power plants. Tests have shown tomatoes grown in such greenhouses contain ten times as much radioactive tritium as the garden variety. While the levels are still low, the dose is of concern due to the particular hazards of tritium which are discussed under the heading Nuclear Power Plants. (*See also*: Phosphate Rock and Fertilizers.)

In both of these cases the food itself is not really radioactive, but it contains small quantities of radioactive substances and can be considered a pathway by which radioactive substances in the environment get into our bodies.

When a radioactive substance leaks into the environment, whether a slight leak due to normal industry operations or more extreme, as it was at Three Mile Island, there is always a risk of these substances finding their way into the food chain and eventually into our bodies.

Radon Gas and Building Materials

Whenever a radioactive substance gives off radiation it undergoes a process known as decay and transforms itself into a different substance. In some cases the new substance formed, the daughter, is itself radioactive; it, in turn, decays into either a stable element or yet another radioactive one. Uranium is at the start of a lengthy chain of these substances. Radon is one of several naturally occurring radioactive daughters created as uranium slowly makes its way down the decay chain to stable nonradioactive lead.

Radon is different from the other daughters (or decay products) because it is a gas and can leak out of the ground, water or building materials and into the air we breathe. When the radon decays into the next element in the chain of daughters or decay products, that element may well be inside someone's lung. There it will continue to decay through the chain, giving off alpha or beta radiation each time it undergoes a step in the decay process.

High concentrations of radon found in uranium mines have long been associated with lung cancer. There is little doubt that radon and radon daughters found in the home are also causing lung cancer, though debate rages over the extent of the problem.

In order to come to grips with the health effects of radon gas, scientists have had to develop a special unit of measurement. Because the harm associated with radon gas is, in large part, due to the presence of radon daughters which cling to dust or moisture in the air and are inhaled, the unit measures the combined concentration of radon and its daughters in the air. It is known as the "Working Level" or WL. To calculate radiation doses, scientists multiply the number of working levels present by the length of exposure time. Since much of the study of radon's health effects has been in occupational settings, the unit of time chosen is usually the number of working hours in an average month, or 170 hours. The dose attributable to one working level of exposure for 170 hours is called a "Working Level Month," or WLM.

The relationship between the damage caused by a WLM compared to the usual dose unit, the rem, is hotly debated. Conversion varies by factors of 200. Accordingly, many researchers prefer to express radon exposure levels only in WLMs and relate the WLM dose to risk estimates directly, without converting to rems first.

The average person is exposed to 0.025 WLMs per year from radon in the air. The average indoor dose has been estimated at one-and-a-half times the outdoor level, or 0.037 WLMs per year. Very high dose rates have been found in homes built over uranium mine tailings in Grand Junction, Colorado, and Elliot Lake, Ontario (where residents are exposed to doses up to 1.5 WLMs per year), and in homes built on top of soil contaminated by waste from nuclear fuel processing plants in Port Hope, Ontario. High levels have also been found in Florida and Newfoundland in homes built with refining slag or over phosphate mine tailings.

The risk of lung cancer associated with radon dose has been calculated by a variety of researchers. Depending on whose numbers you use, a typical exposure to 0.025 WLMs per year leads to a lung cancer risk of between 1.25 and 88 chances in one million for each year of exposure. As much as 8 percent of lung cancer in the U.S. may be due to this natural background radon exposure level. Lung cancer risk due to radon gas is much higher for cigarette smokers (*see* Table 5 below). The pattern of exposure is also important. Unlike most sources of ionizing radiation, the risk per unit of radon appears to be higher if the exposure is slower, according to the supralinear dose-response model rather than the linear model usually employed (*see* The Low-Level Dose Response Debate in Chapter 6).

TABLE 5

Estimated Risk of Lung Cancer for Each WLM

Author	Chances/10,000
Gofman (1981)* (Male Nonsmokers)	3.5
(Male Smokers)	35.3
(Female Nonsmokers)	1.2
(Female Smokers)	11.7
UNSCEAR (1977)†	2.0—4.5
BEIR (1980)†	2.0—6.0
ICRP (1981)†	1.5—4.5
Thomas and McNeil (1982)†	0.5—12.0
AECB—SCRE (1982)†	1.0—6.0

* *Gofman's estimates are for lung cancer death.*
† *As presented by Thomas and McNeil (1982).*

Scientists are now tending to conclude that low doses may be more hazardous than otherwise would be predicted from their only major source of information in this area, data gathered from studies of highly exposed uranium miners (*see* Uranium Miners). On the other hand, the fact that miners are doing physical work which increases their breathing rate may increase their risk, and suggests that there may be less risk for people in other situations.

Factors Affecting Radon Gas Levels in the Home

British surveys have found higher levels of radon in areas characterized by igneous rock formations which are known to contain more uranium. The homes in those areas also have higher radon levels.

High levels of radon have also been found in homes near uranium mines in both Canada and the United States. Most people are not able or willing to move on this basis alone, but those living in uranium-rich locations may wish to take extra care about controlling the different factors which allow radon to accumulate inside their homes.

Three researchers at Princeton University, who noted that radon levels can vary by a factor of 1000 among different homes, have tried to determine the relative importance of the different factors affecting radon gas buildup.

Water supply appears to be one of the greatest potential sources of radon, because approximately half the radon present in household water supplies becomes airborne.

In turn, different sources of water, even in the same geographical area, have different radon levels. Water from lakes and rivers has a much lower radon concentration than water from drilled wells in the same area. Water which is stored in city or town reservoirs also has a lower radon concentration. These two factors point to the greater problems faced by rural dwellers who must often rely on their own wells, and to the advantages of piping water from open bodies of water or from a reservoir. Drinking bottled water will not make a big difference for these dwellers as normally far more radon is released from showers and other wash water in the home.

The Princeton study found that the type of building material used and the rate of ventilation were responsible for far less of the variation between homes than that attributable to differences in water supply. These two factors are nevertheless important and can dominate in cases where the ventilation rate is minimal or where building materials or landfill have come from mine tailings.

Brick masonry or concrete homes are likely to have higher levels than wooden ones due to the radon released from the materials themselves. Also, homes without a poured basement floor allow radon easy access to the home from the soil below. Sealing the concrete in basements with a good-quality paint product can significantly reduce the radon infiltration rate from the concrete and from small leaks in the concrete.

Brick and masonry homes have been found, on average, to have 50 percent more radon than wooden houses. Table 6 shows, however, that there is a considerable range in the content of radioactive substances in common building materials.

TABLE 6

Radioactive Materials

Estimates of concentrations of uranium, thorium and potassium in building materials

Material	Uranium (ppm)	Thorium (ppm)	Potassium (ppm)
Granite	4.7	2.0	4.0
Sandstone	0.45	1.7	1.4
Cement	3.4	5.1	0.8
Limestone Concrete	2.3	2.1	0.3
Sandstone Concrete	0.8	2.1	1.3
Dry Wallboard†	1.0	3.0	0.3
Manufactured Anhydride (by-product gypsum)	13.7	16.1	0.02

† *The estimated concentrations of uranium, thorium and potassium in dry wallboard are those measured by Wollenberg and Smith (1962) for five samples of gypsum which were probably typical of that found in wallboard.*
Source: *Adopted from UNSCEAR (1972).* See Appendix C

A low ventilation rate allows radon to build up and this has led to some concern over the recent emphasis on energy-efficient house construction. Ventilation rate is usually expressed as the number of air changes per hour (AC/hr). A typical house has 0.5 to 1.0 AC/hr in the winter, though summer rates and rates for houses in warm climates are, of course, higher. In older houses ventilation rates are usually between 3.0 and 5.0 AC/hr and some electrically heated homes have a rate as low as 0.3 AC/hr.

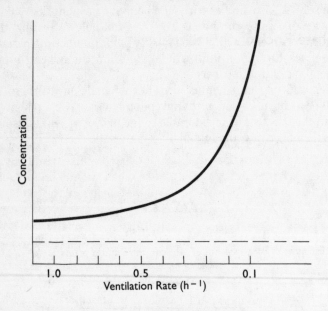

FIGURE 2

Radon concentration and ventilation rate

Source: adapted from H. M. Sachs, T. L. Hernandez, J. W. Ring, "Regional Geology and Radon Variability in Buildings," Environment International, *vol. 8, (1982), Pergamon Press Limited*

As Figure 2 suggests, radon gas levels climb fast when the ventilation rate drops below approximately 0.5 AC/hr. Fortunately newer "air-tight," energy-efficient homes have two features which keep radon levels lower. Most of these homes have a well-sealed basement and a vapor barrier inside the masonry walls which helps keep out radon gas. Very tight homes are also often equipped with air-to-air heat exchangers which increase ventilation without a great loss of energy.

Those building or purchasing energy-efficient homes should look for both these features. Ventilation will also reduce the buildup of other toxic substances in the home.

Phosphate Rock and Fertilizers

Phosphate rock is used widely as a source of phosphorus, one of the three main components of most fertilizers. Unfortunately, phosphate rock often contains a relatively high concentration of uranium and its decay products. Consequently, those who handle the fertilizers and, to a lesser extent, those who handle the fertilized food are externally exposed to high-energy radiation. The uranium also finds its way into the plants and grains fertilized, and we all face added internal radiation exposure when we eat those foods since washing does not remove contamination inside the food (*see* Cigarettes and Tobacco for an example of one particular hazard). In some cases, phosphate is added directly to animal feed as a phosphorus supplement enhancing the mineral and uranium content in meat.

The dose to any individual consumer is relatively low, but the collective dose has been estimated by UNSCEAR as 600,000 person-rems per year (EDE). This exposure can be expected to cause between 60 and 2200 cancer deaths per year worldwide (based upon UNSCEAR's and Gofman's risk estimates, respectively).

Far more serious is the exposure due to the use of building materials, such as phospho-gypsum, a by-product of the mining of phosphate rock. The worldwide effective dose equivalent has been estimated at over 30 million person-rems per year inducing from 3000 to over 100,000 fatal cancers. This is a much higher exposure than that due to most alternative building materials. Individual doses depend on the amount used in a particular dwelling and factors such as ventilation rate and the amount of time spent indoors.

Smoke Detectors

In recent years smoke detectors have become commonplace as millions of the units have been successfully marketed to a public properly concerned about residential fires.

But while there is no doubt that the detectors save many more lives than they threaten, there has been some concern that those brands which contain radioactive substances pose an unnecessary health risk. Most of the estimated forty- to fifty-million smoke detectors in North America have an "ionization chamber" where the reaction between smoke and a radioactive substance—usually americium 241—literally sounds the alarm. A competing technology utilizes a photoelectric cell and contains no radioactive material.

The ionizing detectors can be recognized by a label warning that the devices contain a radioactive substance. They generally contain about one microcurie of americium 241—which emits alpha particles and some gamma radiation—sealed in gold or silver foil inside the air-ionization chamber within the detector.

As long as the substance remains securely inside the detector it is of slight concern. Only the gamma radiation can escape the housing and according to the U.S. Nuclear Regulatory Commission a person who sits ten inches from a detector, eight hours a day for a year would be exposed to less than 0.5 millirems. Under normal circumstances, exposure, if any, would be a tiny fraction of this amount provided care is taken in locating the detectors.

The greatest hazard occurs when the americium escapes due to fire, tampering or damage after disposal. If inhaled it is a potent carcinogen in the lungs. If swallowed it accumulates in the liver, bone marrow and endocrine glands (for example, the thyroid), with a corresponding risk of cancer.

Tests at Oak Ridge National Laboratory in the U.S. have found that when the foil-wrapped radioisotope was exposed to severe fire conditions (925° C for one hour), small amounts of americium escaped (on average 0.3 percent or 0.003 microcuries). The U.S. Nuclear Regulatory Commission has calculated that the maximum amount of airborne americium inhaled or swallowed by a bystander in such a fire would give a dose of 60 millirems to internal tissue. This dose would be spread over the lifetime of the individuals as the americium decayed. Gofman has calculated the americium lung cancer dose (for 25-year-olds) at 0.0128 microcuries. (The lung cancer dose is the dose which is associated with the causation on average of one cancer—whether the dose is received by one person or spread over a great many.) Using Gofman's figure we arrive at a one-in-four possibility of a lung cancer from the escape of 0.31 percent of one microcurie. This presumes that all the escaped americium is eventually inhaled by someone or spread among several people. Ironically, though the detectors make their job easier, the risks are of greatest concern to firefighters who routinely battle residential fires.

But getting rid of a smoke detector is not getting rid of the problem. Neither the U.S. Nuclear Regulatory Commission nor the Canadian Atomic Energy Control Board presently require special disposal techniques for detectors when they are discarded by residential users. As a result, most will end up in municipal dumps where they may be burned.

Taking the americium lung-cancer dose as 0.0128 microcuries, the one microcurie found in the typical detector can cause up to 78 lung cancers if all of it is dispersed and inhaled. With 40 million to 50 million of the units presently in circulation and millions more being produced, the potential risk is growing. The units are expected to last ten years while the half-life of americium is 458 years. Thus, long after a detector has been discarded, the americium lives on, emitting alpha and gamma radiation. Even if only a slight fraction of the americium eventually escapes—say one part in one thousand, and presuming only one in one thousand parts that escape are inhaled, we can expect fifty million detectors to release 50,000 microcuries of americium of which fifty will be inhaled causing 3900 lung cancers in North America. Moreover, this estimate does not include cancers caused by swallowing the americium or the risk of genetic damage. These figures, while highly speculative, are disturbing and lead to serious questions about the use of ionization-type detectors rather than photoelectric detectors.

Both *Consumer Reports* (October 76 and January 77) and *Canadian Consumer* (1975, 1978 and August 1980) have looked at the two types of detectors and come to these conclusions: Photoelectric detectors are better at detecting smouldering or smoky fires (75 percent of all residential fires start this way), while ionization detectors are faster at detecting "smokeless," open-flame fires (the second largest cause of fatalities). As a result, both *Consumer Reports* and *Canadian Consumer* recommend installing both types. (The articles are worthwhile referring to for recommended brands and other pertinent information on installation, etc.)

The ionizing detectors are generally less expensive and if the choice is between a less expensive ionizing unit and nothing at all, it is safer to install the detector. Photoelectric units, while definitely more expensive, appear to be as effective (safer for the more common variety of fire but less sensitive to high-flaming fires) and avoid all risk due to radiation. They may well be the best option.

Static Eliminators

Radioactive substances are often used to reduce static electrical charge in industry, though only two consumer products employ this approach: static eliminators for phonograph records and for photographic negative slides and lenses.

These devices—which usually look like plastic guns—sometimes contain polonium 210, an alpha and gamma radiation source. UNSCEAR has estimated the average annual effective dose equivalent to users due to the gamma rays at 0.01 millirem. This is a relatively minor dose which can be reduced further by holding the device away from your body when using it and storing it in a safe place where it is out of the way of children.

The possibility of the device being damaged by accident or in a fire, and releasing polonium into the air is a serious concern because of the risk of lung cancer if the substance is inhaled. In the United Kingdom, the National Radiological Protection Board has found this to be a reasonable possibility and has estimated that resulting doses could exceed the ICRP limits for public exposure.

Devices utilizing piezo electric technology offer a good alternative to the radioactive variety.

Television Sets and Home Computers

Whenever high-speed electrons are slowed or stopped, as they are in vacuum tubes, X rays are produced. Most but not all tubes in radios and TVs operate at a voltage low enough to limit the power of the X rays produced so that they are contained by the glass tube. Older television sets have three tubes which do emit X rays that can escape the set. Newer solid-state TVs retain only one of these tubes, the picture tube.

Black-and-white sets operate at a lower voltage level than color sets and are not a source of serious concern. Among color sets the older ones are of greatest concern, both due to the two X-ray sources within them which are not found in newer sets and because of the imposition of tighter X-ray emission regulations in the early 1970s.

The regulatory limit is 0.5 milliroentgen per hour measured 5 centimeters (2 inches) from the surface of the set. (A milliroentgen is very close to a millirad.) The actual dose received by the viewer depends upon the time spent watching and the distance from the set. The picture tube size does not appear to be an important factor.

The average annual dose to the gonads has been estimated at between 0.7 and 1.5 millirads for men and 0.2 to 0.4 millirads for woman (lower for women since the X rays are partially blocked by abdominal tissues). The dose can be reduced by shortening viewing time and by sitting farther away from the set.

Recently, however, a new problem has emerged as a result of the growing popularity of home computers and video games. Children (who are likely more sensitive to X rays) are sometimes given the old family color TV to use as a display screen, and they have a tendency to sit very close to the screen. Doses to sensitive organs such as the thyroid could reach 800–900 millirems per year, far above the 100 millirems per year dose limit for children recommended by the National Council for Radiation Protection.

Use of older color sets for this purpose should be avoided, and all viewers should be discouraged from sitting close to color sets.

Newer color sets used as monitors and color computer display terminals are less of a concern than the older TV sets but because they give off more high-energy radiation than one-color (black and white or green and white) screens, they give rise to the same concerns about low-energy radiation as video display terminals (*see* Video Display Terminals for a further discussion of the effects of prolonged exposure to low-energy radiation).

Video-Game Parlors

Video games in public arcades present the same
hazards as televisions, with one serious addition. Some
television screens have been found to leak more high-
and low-energy radiation from the side or back than
from the front. In most arcade games, the user is
looking ahead at a mirror or down at a screen.
In either case, the side of the screen is at waist
level, very near the user's reproductive organs.
Accordingly, the risk of genetic damage is probably
greater from parlor video games than from the home
variety.

Watches and Clocks

The risk of occupational exposure to radiation has seldom been
more tragic than in the case of the many women who suffered
"radium-dial-painter's disease." These young women made their
living by hand painting the dials of watches, clocks and gauges.
They would routinely lick their brushes to keep them pointed and,
in so doing, swallow substantial quantities of radium, the radioactive substance which gave the dials their luminescence.

By 1931 it was clear that these women were suffering a dramatically increased risk of jaw necrosis (jaw-rot as it was known),
bone cancer and aplastic anemia.

Today very few watches are made with radium dials but radioactive substances are still used to provide a luminescent effect.

Watches, clocks and instrument dials which contain such substances can be easily recognized because they glow steadily in the
dark even without ever being exposed to light. Watches which only
glow for a time after being in bright light are not radioactive nor
are those clocks that require electric power to light up or watches
that glow only when a button is depressed.

While radium 226 is now seldom, if ever, used on watch faces,
it is still used on some clock dials.

Other radioactive substances used today are promethium (Pm-147) and tritium-based paints, and gaseous tritium light sources (GTLS) found in small glass tubes inside some LCD watches. (GTLS watches bear a label indicating they contain a radioactive substance.)

The users of these products are exposed to some radiation from the watch or clock directly and some due to swallowing or breathing the radioactive substances which escape into the air.

User Dose Rates

Tritium-Based Painted Dials. Individual internal whole-body user doses of beta radiation have been measured at between 0.06–0.33 mrems per year. The collective dose to wearers in the U.S. has been estimated at 3275–18,012 person-rems per year.

Using Gofman's cancer-induction formula we would expect from 12 to 67 cancer deaths in the U.S. per year from the use of these dials.

GTLS. Intact timepieces containing gaseous tritium light sources give off low-energy X-radiation at a dose rate of less than 0.1 mrad per hour measured 1 cm from the watch. The gonadal dose to a male wearer would be 0.05 mrem per year.

The internal whole-body dose due to routine leakage in homes and offices is estimated at an average of 0.001 mrem for each person in the population each year.

Breaking a tube indoors would expose each person present to between 0.013 and 3.4 rems of beta radiation over eight hours depending on the size of the building and the ventilation rate. While the likelihood of such a breakage is low, the resulting exposure is significant (a 3.4 rem whole-body dose to each of three people would, according to Gofman, result in a risk of one in twenty-seven that one person would develop a fatal cancer and a higher rate if the exposed people are young).

Promethium-147 Painted Dials. PM-147 gives off gamma radiation. Estimates of the whole-body dose rate from watches with PM-147 painted dials range from 0.0028 mrem per year if worn 16 hours per day on the outside of the wrist to 0.21 mrem per year if worn 24 hours per day on the inner wrist. A clock in a bedroom would give an average dose of 0.013 mrem per year to each person exposed.

Radium-226 Painted Dials. Radium gives off comparatively high levels of penetrating gamma radiation. The gonadal dose for males is up to 40 mrem per year, for females up to 24 mrems per year. Doses to the forearm have been found to exceed the 7.5 rem per year regulatory limit by as much as 36 times in some cases.

In the U.S. radium dial clocks have been estimated to give a population exposure of 16,675 person rads per year. Gofman's analysis predicts that 62 cancer deaths per year in the U.S. can be attributed to these clocks.

Other Concerns

Like most consumer products, watches have a limited life expectancy after which, though they may lie unused for many years, most are disposed of. The tritium, promethium and radium in these watches can be expected to find their way from garbage dumps and incinerators into the air and water. Workers who apply the paint or manufacture the tubes will face routine exposures along with the occasional accident and there are corresponding leaks into the environment. All of these exposures—whether from use, manufacture or disposal—will cause cancer and genetic disorders. Any such risk seems unnecessary for the slight benefit offered by the use of radioluminescent dials.

Alternatives such as photoluminescent dials (which glow for a period of minutes or hours after exposure to light) or battery-operated lamps (which light the watch face when a button is pressed) are readily available. Accordingly, use of radioactive watches and clocks is not recommended.

CHAPTER THREE

Radiation

in the Environment

The greatest threats to the health and survival of the human species are those that damage the physical and biological world we depend on to survive.

Radiation sources in the household, workplace or hospital have important implications for the individual and for the society that bears the cost of health care, but these sources of risk seldom jeopardize the health of entire communities, let alone the entire species.

Higher levels of environmental radioactivity, whether from the nuclear power fuel cycle, weapons production and testing, or in the worst case, nuclear war, do pose such a threat—primarily due to the effect of high-energy radiation on the genes. An increased rate of damage to the genes, the biological blueprint for life, can lead to an ever-increasing rate of birth defects and hereditarily linked disease.

No one knows for sure the extent of the problem. Various agencies have tried to estimate the number of people affected by a given increase in a population's exposure to high-energy radiation but all of these estimates are highly speculative (see Genetic Effects). Scientists have yet to agree on which diseases are hereditary or are associated with an hereditary predisposition, but doctors already know of enough genetically linked disorders to routinely ask for family histories as part of diagnosis.

Sources of low-energy radiation in the environment are somewhat less worrisome. Unlike high-energy radioactive substances which can stay in the environment for centuries, low-energy radiation sources, other than natural sources such as the sun, can be "turned off." If we find, for example, an increased incidence of cancer or genetic damage in populations living near transmission lines we, as a society, can decide to control the hazard (although this won't help those already exposed or their offspring). We could choose, for example, to legislate buffer zones near powerful sources and limit unnecessary use of powerful transmitters.

But leaks of high-energy radioactive substances into the environment from uranium mining, nuclear power and weapons production plants seem to be inevitable despite extraordinary measures taken to contain them. And as if the problem of accidental spills were not bad enough, we have a history of regulators permitting planned releases of radioactive substances into the environment from these industries.

The lack of control we as individuals can exercise over our exposure to both high- and low-energy radiation sources in the environment is disturbing. There is no escape from the radioactivity released into the atmosphere by bomb tests, or into our drinking water by nuclear power plants, and individuals have little control over the choice of sites for radio transmitters and electric power lines. Unfortunately the legislators and regulators have not seen fit to ask us for our views on these issues.

As more people learn of the dangers and uncertainties associated with virtually every type of radiation, it is hoped that regulators and governments will have to respond to public pressure by allowing broader participation in plans and decisions involving radiation.

"Natural" Background High-Energy Radiation

Two-thirds of the average person's annual exposure to high-energy (ionizing) radiation is from natural sources. This exposure comes from two broad categories—cosmic and terrestrial sources—and is usually called background radiation.

Cosmic Radiation

Cosmic sources of radiation, such as distant stars, continually expose us to radiation through three mechanisms: creation of radioactive substances in the atmosphere due to neutron bombardment, direct neutron bombardment and radiation produced by the interaction of cosmic particles with the atmosphere.

The prime radionuclides produced in the atmosphere are tritium from which we receive a very low dose of approximately 0.001 millirads annually, and carbon 14 which gives a dose of roughly 1.3 millirads per year.

Direct neutron bombardment, while of greater destructive power, amounts to only 0.35 millirads per year.

The largest dose is due to the radiation other than neutrons which is approximately 28 millirads per year at sea level and higher for locations substantially above sea level (e.g., Denver; 50 millirads per year). Gofman estimates that approximately 14,000 or 3.6 percent of the 389,000 cancer deaths in the U.S. in 1980 were due to cosmic radiation.

Terrestrial Sources of High-Energy Radiation

There are many naturally occurring radionuclides, some which give us an external dose, and some which we inadvertently swallow, thereby receiving an internal exposure.

The largest sources of external radiation are potassium 40 and rubidium 87 and the elements produced from the decay of uranium 238, uranium 235 and thorium 232. The natural incidence of these elements varies from one location to another. Buildings shield their inhabitants from some of the exposure (and from some cosmic rays) but themselves can give off radiation if they contain brick or concrete.

The United Nations Scientific Committee on the Effects of Atomic Radiations estimates the average annual exposure from external terrestrial sources in the U.S. at 32 millirads.

Internal doses are predominantly due to the potassium 40 contained in our bodies, with an average dose to the whole body at approximately 17 millirads.

Other radionuclides provide approximately 1.4 millirads per year.

Naturally occurring radon and thoron gas and the decay products of these gases provide additional exposures although the doses cannot be expressed in rads or rems.

Gofman has calculated that for an equilibrium population of 250 million (a population neither growing nor shrinking for many generations) the total terrestrial dose would produce 55,500 cancer deaths per year. He estimates 28,000 deaths per year currently in the U.S. can be attributed to this cause.

Background Radiation Levels as a Safety Guideline

Clearly a very large proportion of our exposure to ionizing radiation is from natural sources and is, therefore, largely beyond our control. But the fact that it is natural and uncontrollable does not make it safe, and should not lessen our concern over the radiation sources we do control.

Comparisons of radiation levels from man-made sources to the natural background level are often made by people in radiation-related industries to dismiss concern over additional exposure risks, and are of little help. Such estimates imply that the background level is risk free, which it is not. Every additional bit of exposure represents an additional risk and cannot be justified on the ground that we already bear some risk.

Sunlight

Sun worshippers celebrate a star which is essential to life on earth. But, as with so many good things—too much sunlight can be a hazard.

Sunlight is composed of a wide range of frequencies, or energy levels, of electromagnetic radiation. Most of these frequencies (or wavelengths) are harmless given the amount we receive, but some—in the higher energy invisible ultraviolet range—are associated with skin cancer.

According to experts, sunlight may cause 90 percent of all skin cancers. Skin cancer incidence in the United States ranges from 300,000 to 400,000 cases per year and is responsible for approximately 10,000 deaths annually. Sunlight is also associated with the wrinkling and mottling due to loss of elasticity which prematurely ages the skin.

There is also some evidence that cataract formation is associated with sunlight. Since sunglasses do not filter out all the ultraviolet light, wearing them may even increase eye exposure because they suppress the normal protective measure of squinting.

For skin cancer due to sunlight there appears to be a threshold effect: the risk of cancer increases dramatically above a certain level of exposure. In other words, a slight tan may act as a safeguard but excessive tanning heightens the risk. Such an effect may be due to the protective action of a slight tan and could help explain the higher incidence rate of skin cancer among fair-skinned people who don't tan as easily.

Precautions are largely a matter of common sense. Stay out of the sun when its rays are most direct—midday and midsummer—and wear sun lotion with a screening agent if you will be receiving a lengthy or sudden exposure. Tanning parlors which offer tanning under artificial light may be more harmful than natural sunlight due to the makeup of the lamp spectrums (*see* Fluorescent Lighting).

Long-term societal preventative measures include the reduction of aircraft exhausts and chlorofluorocarbons. It has been predicted that a buildup in the atmosphere of these substances which allow increased transmission of ultraviolet light may increase the incidence of skin cancer by 60 percent or more in the early twenty-first century.

Microwaves and Radio Frequency (RF) Radiation

In the mid-sixties readers of the morning paper were treated to a news story with all the intrigue and drama of an Ian Fleming spy thriller: The American Embassy in Moscow was being showered with microwave radiation by its Soviet host. The "Moscow Signal," as the beam became known, puzzled U.S. intelligence agencies as it did not appear to be appropriate for either electronic eavesdropping or for jamming any electronic eavesdropping devices that the embassy might have been using. A review of published Soviet and East European research led to speculation that the beam may have been an attempt to disrupt the behavior of embassy personnel. Eventually the embassy erected metal screens to shield the building's occupants from the waves.

The energy level of the Moscow Signal was about 4 milliwatts per square centimeter ($4mW/cm^2$). Then, as now, the permissible rate of exposure to microwave and radio frequency (RF) radiation in the U.S. is 2.5 times the power of the "Moscow Signal," or 10 mW/cm^2. The Canadian standard was tightened from 10 down to 1 of these units in 1979.

This discrepancy between the permissible exposure limit and the protective actions taken by the embassy when levels were less than half the limit is not only of interest to those who follow the

FIGURE 3
Microwave transmission tower

progress of the "Cold War." We are all bathed in radio frequency (RF) and microwave radiation, virtually continuously, and the embassy's response indicates that we might reasonably be concerned about limiting our exposure.

Yet, these two types of radiation (actually a continuous range of frequencies with an ill-defined line dividing it into the two categories) are widely used for television and radio transmission, telecommunications, airport and police radar, satellite communications, burglar alarms, garage-door openers, medical heat treatments (diathermy), industrial heating and drying processes and in microwave ovens.

The greatest exposures are for the occupationally exposed and as Table 7 indicates the use of microwaves and RF sources in industry is widespread.

Microwaves and RF waves are low-energy forms of radiation but, like all radiation, exposure can be serious if the dose is high enough. And as with most other forms of radiation the scientific community does not have one voice telling us where, if anywhere, a safe level lies.

At high levels such as those in microwave ovens, the energy levels can literally cook tissue. It is this effect which led to the 10 mW/cm^2 safety limit, as tissue heating has been observed above that level. Below 10 units, our biological cooling systems can keep up and protect most parts of the body from the excess heat energy before damage or discomfort is apparent. The two areas of the body which may not be so well protected are the eyes and the brain and central nervous system.

A number of researchers have described the observed or likely effects of prolonged exposure at different energy levels.

Low doses (less than 1 mW/cm^2)	• heating effects not apparent
	• possibility of headaches, irritability, sleep disturbance, weakness, decrease in sexual activity and chest-pain
Medium doses ($1–10 \text{ mW/cm}^2$)	• increased effects similar to those listed above as well as

fatigue and non-permanent
changes in the E.E.G. measurements of nervous system
functioning and in E.C.G.
measurements of cardio-
vascular system functioning
• lowering of blood pressure

High doses   • severity of effects increases,
(above 10 mW/cm²)   acute crisis of the circulatory
system and thermoregulatory
disturbance

In addition to the above effects, microwaves can cause cataracts in the eyes. The energy level and length of exposure time
required to cause cataracts has been a subject of some debate with
much of the North American research being conducted and controlled by U.S. government agencies who claim secrecy privileges
based upon national security. Skeptical commentators have expressed the opinion that the "national security interest" which has
limited research or publication of findings has been most inter-
ested in keeping the thousands of technicians and armed services
personnel who are occupationally exposed to RF and microwaves
in the dark and out of court.

Some researchers feel a one-hour exposure to 100 units is
required to form a cataract, but there is evidence that even a
prolonged 2-unit occupational exposure can induce the irregulari-
ties in eye tissue.

Researchers emphasize that the data on health effects which
have been collected so far are for exposure of healthy adults, mostly
males. The effects on children are unexplored and differences in
body size may well lead to a different pattern of energy absorption
and effect. Heightened effects on unhealthy or otherwise sensitive
individuals would not be surprising.

There is very little reported research on the effects of microwave and RF radiation on children in the womb or on the genes
which could affect the health of offspring. One Johns Hopkins
University study has found a significantly higher number of chro-
mosomal abnormalities in the blood of men exposed to radar in

their jobs than in unexposed workers. An earlier study also conducted by Johns Hopkins had found that fathers of Down's syndrome children (children afflicted with mongolism) were three times as likely to have been exposed to radar as the fathers of unaffected children in the study. Two recent findings, showing that the genetic material DNA absorbs microwave radiation readily, and that DNA changes are accompanied by a molecular vibration at a frequency near that of microwaves, are indicative of a possible connection between microwaves and genetic damage.

At present there is no evidence of an increased rate of birth defects in the general population living in areas where routine exposures due to radio or television transmissions are higher. This absence of evidence may be because no significant effect exists at the energy levels experienced in those areas or it may be simply because no one has taken a careful look. Russian scientists are apparently concerned— the public exposure limit in Russia is one-thousandth that permitted in the U.S. and one-hundredth of the Canadian limit.

Exposure is higher the closer you get to the source as indicated in Table 8. Those working on tall buildings may be exposed to emissions from transmitters on top of their own or nearby buildings.

Microwave and RF:
Guidelines *versus* reality

NIOSH, the U.S. federal government agency responsible for investigating occupational health and safety issues, has reported that of 1,400 measurements of various RF sources typically found in the workplace, at least 75 percent of the sources exceeded the personnel radiation protection guidelines set by the standards setting agency ANSI.

from: "Documenting Radiation Effects" (a report on work at NIOSH) in, Occupational Health and Safety, *October 1979.*

TABLE 7

Selected examples of the typical uses of equipment generating radio frequency and microwave radiation

Frequency	Use	Occupational exposure
Below 3 MHz	Metallurgy: eddy current melting, tempering; broadcasting, radiocommunications, radionavigation	Metal workers; radiotransmitter personnel.
3–30 MHz	Many industries such as the car, wood, chemical and food industries for heating, drying, welding, gluing, polymerization, and sterilization of dielectrics; agriculture; food processing; medicine; radio astronomy; broadcasting.	Various factory workers, e.g., furniture veneering operators, plastic sealer operators, drug & food sterilizers, car industry workers; medical personnel; broadcasting transmitter and television personnel.
30–300 MHz	Many industries, as above; medicine; broadcasting, television, air traffic control, radar radionavigation.	Various factory workers, as above; medical personnel; broadcasting transmitter and television personnel.
300–3000 MHz	TV, radar (troposcatter and meteorological); microwave point-to-point; telecommunication telemetry; medicine; microwave ovens; food industry; plastic preheating.	Microwave testers; diathermy and microwave diathermy operators and maintenance workers; medical personnel; broadcasting transmitter and television personnel; electronic engineers and technicians; air crews; missile launchers; radar mechanics and operators and maintenance workers; food industry workers.

Frequency	Use	Occupational exposure
3–30 GHz	Altimeters; air- and ship-borne radar; navigation; satellite communication, microwave point-to-point.	Scientists including physicists; microwave development workers; radar operators; marine and coastguard personnel; sailors, fishermen and persons working on board ships.
30–300 GHz	Radiometeorology; space research; nuclear physics and techniques; radio spectroscopy.	Scientists including physicists; microwave development workers; radar operators.

Source: ENVIRONMENTAL HEALTH CRITERIA 16, Radiofrequency and Microwaves, World Health Organization, Geneva (1981)

TABLE 8

Power Density at Various Distances from a 50,000 watt
AM Radio Station

Distance (feet)	Power Density (μW/cm^2)
15	838
29	284
69	196
152	43
308	33
482	23
663	12
1571	2
3280	1
5760	0.3

Source: Marino and Becker. See Appendix C

TABLE 9

EMF in Typical Tall Buildings

City	Location	Power Density (μW/cm^2)
New York	102nd Floor, Empire State Building	32.5
Miami	38th Floor, One Biscayne Tower	98.6
Chicago	50th Floor, Sears Bldg.	65.9
Houston	47th Floor, 1100 Milam Building	67.4
San Diego	Roof, Home Tower	180.3
Toronto	CN Tower (Public Area) (Worker Area)	up to 300 up to 20,000

Source: Adapted from Marino and Becker. See Appendix C

High-Voltage and House-Current ELF Radiation

The alternating current (AC) which travels along high-voltage transmission lines and eventually finds its way through a wall socket into the reading light next to your favorite chair is the source of two types of low-energy radiation, electric and magnetic, both of which may have subtle but, in the long term, serious health implications.

Electricity in North America is transmitted through wires at a standard frequency of 60 cycles per second (60 Hz). In Europe the standard is 50 cycles per second. This pattern of an alternating polarity pulse allows for simplified voltage changeovers and is useful for a number of electrical applications. However, alternating-current electricity transmission also floods our environment with magnetic and electric fields which fluctuate at the same characteristic frequency of 60 cycles per second, a frequency in the ELF (extremely low-frequency) range.

Scientists have noted a variety of biological effects in animals and humans from laboratory-controlled exposure to such fields. These effects are usually subtle physiological or behavioral changes and most researchers are not ready, given the limited research done, to state that the effects are harmful.

Unfortunately, the silence of the researchers is far from reassuring in this particular area, since the great majority of ELF research is funded or controlled by organizations such as the utility-owned Electric Power Research Institute or by the U.S. Navy which have strong vested interests in proving that ELF isn't dangerous.

Accordingly, researchers have for the most part managed to avoid studies of the effect on humans, concentrating instead on animal studies which offer less valuable information to those concerned with human health effects. One industry-funded study which did look at electric-utility linemen found two of the eleven men examined had low sperm counts.

The scientists at Johns Hopkins University who did this study asked for funding to do further work in the area but they were turned down by both industry and government.

The U.S. Navy's interest in ELF radiation stems from its mid-1960s proposal to build a massive ELF transmission system

FIGURE 4
A microwave communications tower amidst a maze of electric power
switches and transformers—microwaves are often used by power
companies for communication.

FIGURE 5

High-voltage switching and transformer stations are often located in residential areas.

buried throughout 26 counties in northern Wisconsin to enable communication with underwater submarines.

The project, known first as Seafarer and later as Sanguine, came under the U.S. National Environmental Policy Act which requires an assessment of environmental impact. The U.S. Navy was forced to fund a great deal of ELF research most of which resulted in findings of biological effects.

The Sanguine research did not, however, focus on negative human health effects.

From the research that has been done on human health there is no reason to conclude that the biological effects are not harmful. Findings include:

- lowered sperm counts among linemen;
- possible neurological and cardiovascular disorders in Russian high-voltage switchyard workers;
- fewer children born to exposed high-voltage workers than to their unexposed colleagues and an increasing difference with increased years of exposure.

All of these findings are for high-voltage field exposures found in industrial settings and, given the small populations studied, must be viewed as tentative.

The findings are nevertheless of concern to those who work or live near high-voltage power lines for the fields around such lines are measurable from thousands of feet away.

There have been measurable effects found in animals exposed to weak fields equivalent to those found one mile away from a high-voltage transmission line. Altered human physiology or behavior has been found in experiments where people have been exposed to field strengths similar to those experienced 1400 feet away from a 765,000-volt transmission line commonly found in North America.

More recent research by independently funded scientists has revealed some disturbing relationships between ELF magnetic-field exposure and negative health effects. In 1979 the *American Journal of Epidemiology* published a report indicating a relationship between electrical wiring configurations and incidence of childhood cancer. The study suggests that the magnetic field associated with high-current flow (as opposed to higher voltage) may be the culprit. Children with cancer were more likely to be living near high-current flow transformer stations.

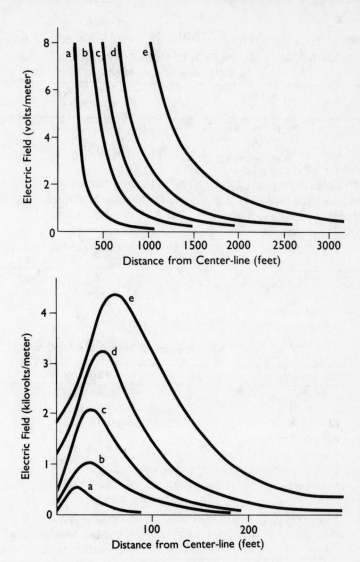

FIGURE 6

Ground-level electric fields of typical high-voltage power lines.

a, 115 kv; b, 230 kv; c, 345 kv; d, 500 kv, e, 765 kv.

Source: Electromagnetism and Life, Robert O. Becker and Andrew A. Marino. See Appendix C

TABLE 10

Power-frequency Electric Fields of Household Appliances Measured at a Distance of One Foot

Appliance	Electric Field (v/m)
Electric blanket	250
Broiler	130
Phonograph	90
Refrigerator	60
Food mixer	50
Hairdryer	40
Color TV	30
Vacuum cleaner	16
Electric range	4
Light bulb	2

Source: Marino and Becker. See Appendix C

TABLE 11

Power-frequency Magnetic Fields of Household Appliances

Range	Appliance
10–25 gauss	Soldering gun Hairdryer
5–10 gauss	Can opener Electric shaver Kitchen range
1–5 gauss	Food mixer TV
0.1–1.0 gauss	Clothes dryer Vacuum cleaner Heating pad
0.01–0.1 gauss	Lamp Electric iron Dishwasher
0.001–0.1 gauss	Refrigerator

Source: Marino and Becker. See Appendix C

Another study found a high incidence of acute leukemias among workers exposed to high-voltage and high-strength magnetic fields in their work. The study looked at 438,000 deaths among working men in Washington and found that the exposed group was two-and-one-half times more likely to suffer acute leukemias than the unexposed group.

A third study which focused on magnetic fields found in residential situations uncovered a relationship between field strength and suicides.

Finally, an ongoing study of 450,000 Canadian male workers has found that telephone, telegraph and power line workers run three times the average risk of dying of leukemia.

While scientists know a lot more about ELF now than they did ten years ago, they are almost totally in the dark about the mechanism of its effects. Scientists have not been able to isolate physical or chemical interaction which would cause the effects and which, once known, could be countered by protective measures. The relationship between electric and magnetic fields, both of which are usually present near power lines, and the subtle electrochemical processes of the body is simply not understood.

Until the researchers have isolated the mechanism of biological interaction and have adequately confirmed the level of effect on all population groups through epidemiological studies the risk of ELF radiation exposure will be difficult to quantify. Those living or working near high-strength fields over long periods of time are presumably at the greatest risk.

ELF Theft and Weapons Detectors

It's not uncommon to see detector gates near department store and library exits and at departure gates in airports. Most of these devices produce electromagnetic fields in the ELF range with a strong magnetic component. While unlikely to affect the person being scanned (given the short period of exposure), employees who are stationed near the devices may be wise to keep their distance, if possible.

Nuclear Power

There are now over 300 nuclear power plants in the world (approximately 75 in the U.S. and 20 in Canada), and one large plant produces each year as much long-lived radioactivity as one thousand Hiroshima bombs. With numbers like these it's easy to see that we can't afford mistakes. Just one of the world's reactors leaking 1 percent of its radioactivity into the environment could release as much radiation as 10 bombs.

The sad truth is that even without such a catastrophic accident, nuclear power plants are releasing vast amounts of long-lived radioactivity into the environment.

Even without the military use of nuclear power technology, President Dwight D. Eisenhower's "peaceful atom" is wreaking havoc on us and on our descendants.

The Nuclear Fuel Cycle: An Overview

Any debate about the safety of nuclear power is, at its heart, a debate about two issues: the amount of radiation released into the environment by the nuclear fuel cycle and the harmfulness of that radiation to our health. These questions are, of course, similar to those posed in any situation where radiation is a potential problem, but in the case of nuclear power, the public debate has been a prolonged and highly charged one for several reasons:

First, the risk to the public from nuclear power is an imposed risk; it is not voluntarily assumed. Second, the nuclear power fuel cycle is a constant source of long-lasting radioactive contamination of our environment due to routine emissions of radiation and radioactive substances, and the everpresent risk of a catastrophic accident. And third, the benefit provided by nuclear power, i.e., electricity, is one that can be obtained by other means which do not have these kinds of drawbacks. In other words the risk is an unnecessary one and can be avoided without the loss of any benefits. Of course there are other important concerns with nuclear power including its connection with nuclear weapons development that are not dealt with here.

Proponents of nuclear power intent on belittling these concerns often cite data on the rate of release of radiation to the environment by nuclear power plants which conveniently avoids taking account of either the possibility of an accident or the routine emissions due to the steps in the process before and after the fuel is "burned" at the power plant.

Uranium, the fuel used in most nuclear reactors, is a naturally occurring substance which is ordinarily buried in rock deep underground. The greatest sources of radioactivity associated with nuclear power are the mines and mills, where 85 percent of the radioactive substances brought up from below the surface are left lying about in "uranium mine tailings," piles of waste that leak radioactive substances into the air and water (*see* Uranium Mining and Milling).

Next comes the enrichment process which separates the different types of uranium, a step which is necessary for the nuclear chain reaction to occur in light-water reactors, such as those used in the U.S. Here again radioactive wastes are left behind as tailings.

The fuel, either enriched or natural uranium, is then packed into fuel rods and bundles at another location and shipped to nuclear generating stations.

Unlike many hazardous substances the fuel does not become less of a concern once "burned." The "spent" fuel from a nuclear power plant is millions of times more hazardous than the uranium going in. This is so due to the production within the nuclear reactors of "fission products," substances which are produced when the uranium atoms are split in the chain reaction, and substances produced due to the effects of neutron radiation.

One such substance is plutonium 239. It is considered to be one of the most toxic substances known to man, and it is almost entirely man-made. Once formed, we can't get rid of it—an atom of plutonium 239 has an average life expectancy of 35,209 years.

At this stage the spent fuel is either stored for eventual disposal, a growing problem since no safe disposal technique has yet been found, or it is reprocessed to extract the unburned fuel and to separate out plutonium for use in nuclear weapons.

Reprocessing adds yet another step where radioactive substances are routinely leaked into the environment and does not reduce the need for waste disposal of the fission products. At present there are no commercial reprocessing plants in North America, only military ones.

FIGURE 7
The Pickering Nuclear Generating Station on the outskirts of Toronto comprises eight reactors.

In addition to radiation from the fuel itself throughout the cycle, the neutrons set free in the chain reaction cause the reactors themselves to become radioactive and also create radioactive gases and water which leak into the air we breathe and water we drink.

Routine Releases of Radioactivity from Nuclear-Power Production and Associated Radiation Doses to the Public

Even if we leave aside the risks due to catastrophic accidents, determining the risk to public health posed by nuclear power is anything but straightforward.

First we must know the quantity and types of radioactive substances released at each stage of the nuclear fuel cycle. Second, we must estimate to what extent we will actually come in contact with these substances in the air we breathe, the water we drink and the food we eat. We must also take into account the varying degrees of radioactivity and longevity of the different substances.

Finally, once a calculation is made of the likely dose of radiation the public will receive, we must determine the health risk associated with that dose.

Scientists and researchers have made differing estimates and reached diverse conclusions for each of these factors but even the most optimistic assessments are disturbing.

UNSCEAR has estimated that for each gigawatt year of electricity produced by nuclear generation the public receives the equivalent of 570 person-rems (effective dose equivalent) within roughly two years and approximately 400,000 person-rems spread over millennia due to the long-lived radioactive substances released. In 1979 there were 55.48 gigawatt years of nuclear-powered electricity generated in the U.S. and Canada. Based on UNSCEAR's assumptions, the resulting dose of radiation was, from that one year, 31,624 person-rems, and mankind will receive (if it survives far into the future) another 22,192,000 person-rems due to nuclear power production in 1979 alone.

If we use UNSCEAR's estimate of 100 cancer deaths per million person-rems, we have sentenced roughly 2222 people to premature death due to cancer from that one year of nuclear power generation in North America. If we use Gofman's estimate of 3771

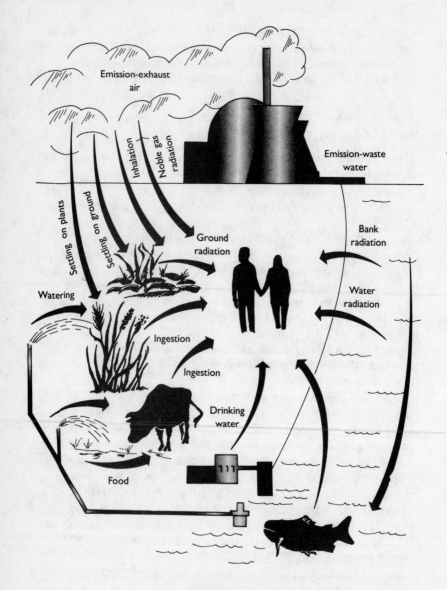

FIGURE 8

Pathways of Exposure
from Nuclear Power Plants

TABLE 12

Estimates of the Collective Radiation Dose to
the Public from Each Year of Nuclear Power Production

	Person-Rems
Local and Regional (Short-term)	
Mining (Radon)	4000
Milling (not including releases from mine tailings) (Uranium, Thorium, Radium, Radon)	320
Fuel Fabrication (Uranium)	16
Reactor Releases	
Airborne (Noble gases, Tritium, Carbon-14, iodines and Particulates— Cesium, Ruthenium, Cobalt)	32,800
Waterborne (Tritium, Cesium, Ruthenium, Cobalt 60)	480
Fuel Reprocessing	
Airborne Releases	2400
Waterborne Releases	5600
Transportation of Radioactive Substances	24
Global (Long-term)	
Tritium	160
Krypton 85	15,200
Carbon 14	880,000
Iodine 129	4,480,000
Mill Tailings	
Radon	22,400,000
Uranium	3,680,000
High Level Wastes	240,000
Total	31,741,000

NOTE: *Based on 1980 nuclear power generation of 80 gigawatts, partial listing only of substances emitted, does not include doses due to serious accidents. All figures must be considered rough approximations, especially long-term estimates which are based on optimistic assumptions about containment of tailings and high level wastes.*
Source: UNSCEAR (1982). See Appendix C

deaths per million person-rads we can expect 83,805 cancer deaths. Most of these deaths will be very far in the future, but a few will be among those alive today (from 3 to 120 depending on whose estimate you accept).

These figures do not account for the suffering future generations will face due to genetic damage. Two other factors should also be considered: In 1985 we are producing even more nuclear power than was produced in 1979, and in the coming years there is every reason to believe that a serious reactor accident is a real likelihood. Such an accident would have both immediate and long-lived consequences. To make matters worse the UNSCEAR estimate is based on an optimistic view of the reliability of (an as yet to be determined) long-term, high-level waste disposal technique and does not include the radiation exposures of those people employed in the various mines, mills and plants (*see* Uranium Miners and Radiation Workers).

Nuclear industry apologists who claim there has never been a death attributable to nuclear power conveniently forget about lung cancer among uranium miners and the thousands of cancer deaths in the general population that have been predicted by even the most optimistic of scientific bodies studying the issue.

From Reactor Releases to Actual Doses: More Than We Bargained For?

In determining how much of a given radioactive substance a nuclear plant is allowed to release to the environment, plant operators and regulators are supposed to evaluate the amount that will find its way into our bodies and the radiation dose that will result. Such a calculation is not a simple matter.

Each stage of the food chain must be examined to determine how much of each radioactive substance gets absorbed by each type of plant or animal, and how much gets passed along to humans. Different factors must be considered for airborne, as opposed to waterborne, releases. Weather patterns and population distribution must also be considered.

For years, the U.S. Nuclear Regulatory

Commission has used a model for these calculations which is based upon experiments conducted in the 1950s by Atomic Energy Commission scientists who sought to reassure the public about radiation doses from bomb testing. In 1979, fourteen University of Heidelberg scientists released a study demonstrating that the AEC scientists had rigged some of their results and that those living near nuclear plants are receiving doses far higher than we had been led to believe. The study suggests, for example, that the NRC's evaluation of the amount of plutonium, cesium and strontium crops will pick up from the soil is from ten to one thousand times too low.

In 1982 Bernd Franke, a scientist with the Heidelberg IFEU Institute, presented a paper demonstrating further scientific weaknesses in the methodology used by the NRC. He pointed out that transfer factors—the numbers which represent the fractions of radionuclides that are transmitted from one step in the food chain to the next (e.g., from the soil to plants or from crops to milk)—were derived from a variety of inappropriate data sources which have little relationship to actual situations. One example is the transfer factor for fodder to meat which is based on mixed data for radionuclide levels in American towns, English plants and Russian soil and not on consistent measurements near an actual power plant in the U.S., or anywhere else for that matter.

Canadian limits are similarly based upon mathematical constructions using data borrowed from diverse sources rather than from actual experience.

Franke goes on to point out that the transfer factors adopted by the NRC for such critical substances as strontium are markedly lower than those found by most scientists. The NRC disputes the Heidelberg findings but is not willing to put the matter to a test that Franke proposes: calculations based upon a series of real measurements near the various nuclear power plant sites. Perhaps they fear what they would find.

TABLE 13

Summary of Estimates of Fatal Cancers Due to One Year
of Nuclear Power Production at 1980 Level
(Excluding Nuclear Workers)

Source	Estimate of Fatal Cancer Rate Per Million Person-rems	Cancers based on UNSCEAR Estimate of 31.74 Million Person-rems Released in 1980
UNSCEAR	100	3174
Beir	70–353	2222–11,204
Gofman	3771	119,692

NOTE: UNSCEAR (1982) estimates of collective effective dose equivalent commitments to the public for 80 GW of power (1980 level) include both present and future generations; do not include deaths due to serious accidents or genetic mutation. (Currently ordered or partially constructed plants will result in a doubling of nuclear power capacity and, therefore, a doubling of these estimates.)

Risk Due to Accidents

Since the accident at Three Mile Island in March of 1979, it has become absolutely clear that the possibility of a serious mishap at any nuclear power plant cannot and should not be discounted.

Scientists had always known that the contents of a reactor are something which must not escape to the environment. Official agencies had calmed anxious members of the public by pointing to the "defence in depth" technology that had been developed to insure the containment of these hazardous substances. However, the lesson of Three Mile Island has shown us that no technology can provide an absolute guarantee when we ultimately rely on human judgment, subject to human error, for the design and operation of the technologies.

In an academic setting, an engineering student may be considered brilliant if he or she can score 99 out of a 100 on an exam, but in the real world of nuclear reactor engineering that one mistake can be fatal.

A number of studies have tried to determine the likelihood of such an error and what the potential consequences of a serious accident could be. As in the case of estimates of risk due to the routine release of radioactivity even the optimistic official estimates are disturbing.

The most recent study was done by Sandia National Labs for the U.S. Nuclear Regulatory Commission. It analyses the consequences of an accident for each U.S. reactor site. The results are also applicable to other countries, and in Canada a Royal Commission of Inquiry has found that there is no reason to believe an accident is any more or less likely to occur at a Canadian nuclear installation than at a U.S. plant.

The Sandia study, which became public in late 1982, gives a range of likelihood both for the occurrence of a serious accident and for the human health consequences (which vary depending on the weather conditions at the time of the leak and the population density near the reactor).

The probability of a large release of radioactivity was found to be between approximately 1 in 10,000 and 1 in 100,000 per reactor per year.

The Union of Concerned Scientists has calculated that these numbers translate into a risk of between 2.6 and 23.5 percent that a very serious accident will occur in the U.S. by the year 2000, given the number of reactors expected to be running over the next 16 years.

According to the Sandia study, the average serious accident would cause from 1 to 970 immediate fatalities, 4 to 3600 injuries and 230 to 8100 cancer fatalities. But the Sandia study goes on to predict the worst-case consequences of such an accident, which are between 173 and 102,000 fatalities, 3130 and 610,000 injuries and 3180 to 40,000 cancer deaths.

Apparently the Nuclear Regulatory Commission is aware of the disturbing nature of these findings. The results were not made public until they were formally requested by a House sub-committee, and the report that was released analysed only the average figures, and buried the worst-case results in the accompanying computer printouts.

These estimates all ignore the added risk of "external" events such as the possibility of sabotage, earthquakes, hurricanes, tornadoes or aircraft crashes which might cause a serious leak of radioactivity into the environment. In the opinion of the authors of a 1983 report to the Union of Concerned Scientists, the study did not adequately deal with human error. Thus, the figures err on the side of optimism and must be read as underestimates of risk. Furthermore, the cancer figures are based on estimates of radiation carcenogenicity which are probably too low and include only "first-generation" consequences, not genetic damage to future generations.

Risks of Living Near a Nuclear Plant

While it should be clear from the preceding sections that the risks associated with nuclear power are spread far wider than the immediate vicinity of the power plant, it is also clear that those living near the plants do face a higher risk simply because the radioactive substances routinely released (and those released in accidents) are present in higher concentrations near the plants.

The Sandia Labs study for the Nuclear Regulatory Commission indicates that the number of people who will develop cancer following a serious reactor accident does not vary a great deal from one reactor to the next. This is because the predominant effects will be to those within 100–200 miles of the plant, and

---

No one Died at Three Mile Island?

Debate over the death toll due to the Three Mile Island accident in 1979 will probably continue for years to come mainly because there is no clear consensus on the type and amount of radiation that leaked to the environment. What *is* clear is that the cleanup after the accident is going to expose workers to between 13,000 and 46,000 person-rems of radiation. Even the U.S. Nuclear Regulatory Commission admits the dose will cause from 2 to 6 cancer deaths and will result in 3 to 12 birth defects among children born to the workers.

---

studies show that the population within 200 miles of any given plant is roughly the same. However, the number of injuries and immediate deaths does vary significantly depending on local population density (within about 25 miles)—indicating that, in the event of a serious accident, the cancer risk is more dispersed than the risk of acute, possibly fatal, radiation poisoning.

When considering routine or accidental releases it is important to distinguish between the dose due to gamma rays, which escape the reactor core to irradiate the immediate neighborhood, and the dose due to the release of radionuclides which can travel in the air, water or in the food we eat for some time and distance before exposing someone to radiation.

The gamma dose just outside the fence surrounding a nuclear plant is the dose which the operators of the plants like to quote when they assure the public of their safety. Because the dose to workers inside the plant must be kept within regulatory limits it is not surprising that the dose outside the plant is relatively minimal, about 1 millirem per year.

But these same operators usually fail to mention the far more serious dose received by inhaling or swallowing the radioactive particles, gases and liquids which routinely leak into the environment from the plants. The dose levels associated with these releases decrease as one moves farther from the plant. As Table 14 shows, average individual dose rates from gases are roughly twice as high within 2 kilometers of the plant as they are for those living 2–5 kilometers from the plant, and five times as high as experienced by those living between 5 and 10 kilometers away.

It is virtually impossible to estimate the dose any individual actually receives because the radioactive substances follow many different paths to the body (*see* Figure 8). Much of the dose is taken in with food and an individual dose will depend on where you buy your food, and what type of food and how much of it you eat. Iodine 131, one radioactive substance released both routinely and during accidents, and which presents a serious thyroid cancer risk, easily enters the food chain when it settles on ground where cows graze and subsequently turns up in cows' milk. Once in the milk, it may be transported several hundred miles away to market or, if made into cheese, some may end up being exported to another continent.

TABLE 14

Dose due to Routine Release of Radioactive Noble Gases from a
Nuclear Power Plant Dependent Upon Distance from the Plant

Distance (km)	Percentage of Individual Dose Compared to Those Living Within 2km (1.25 miles) of Plant
2–5	46
5–10	19
10–20	7
20–50	1
50–100	1/2

NOTE: *UNSCEAR data for noble gas dose commitments due to releases from typical pressurized water reactor.*

There are dozens of different radionuclides released, each one in different amounts, by different pathways and with differing health effects.

The radionuclides released give off either alpha, beta or gamma radiation, or a combination. Some tend to accumulate in specific organs such as the thyroid or the bones while others stay in the lungs. Wherever they end up, radionuclides create health risks concentrated in the part of the body where the substance accumulates.

Apart from moving away from the vicinity of a nuclear plant, there is very little that can be done by an individual to reduce his or her exposure due to routine or accidental emissions.

Local residents should, however, insist on being kept informed on the rate of release of radioactive substances from the nuclear facility to be sure that the situation is within legal limits and that it is not getting worse.

Accidental releases (even relatively minor ones) usually involve a higher concentration of radioactive substances in the air or water for a period of hours or days. There are measures available to reduce exposures from such accidents. For example, in the event of a serious accident, such as the Three Mile Island case, residents would be wise to leave the area for several days, especially children and pregnant women.

Some communities near power plants have been provided with iodide pills which, if taken at the first sign of an accident, saturate the thyroid gland with iodine and thereby reduce the absorption of radioactive iodine over the following few hours. If you live near a plant, and have not been provided with these pills, consider contacting the local authorities to encourage the distribution of potassium iodide pills as a precautionary measure. Distribution of the pills following an accident, even if practical, is of much less value because you may have already absorbed radioactive iodine.

If you learn of an accidental radiation release, contact the local health authorities to determine the amount of radiation released and the pathway. If the pathway is via the water you may wish to drink bottled water for several days—especially if you are pregnant. This is a particular concern for Canadians living near Candu reactors because of the buildup of the radioactive substance tritium in the water used to cool these reactors. There have been numerous spills of tritium-contaminated water into the Great Lakes water system. Like other long-lived radionuclides, tritium persists in the environment where some of it finds its way into human bodies. Because our bodies cannot distinguish between tritium–contaminated water and normal water, and the substance can be incorporated into virtually any part of the body or, by crossing the placenta, into the body of an unborn child, it presents a serious hazard.

There have been thousands of what the nuclear industry labels "minor accidental releases" of radioactivity into the environment but, to date, only two have been labeled serious: the 1957 Windscale fire in England and the incident at Three Mile Island.

The Windscale fire is suspected to have caused several hundred cancers, particularly thyroid cancer. Recently there have been indications that the leukemia rate among persons under 25 years of age in the vicinity of the plant (now renamed Sellafield) may be four times the British average. This elevated leukemia rate could be due to the accident or to the routine emissions from the fuel reprocessing which is done at the plant. Recently the public was urged to stay away for several weeks from an 8-kilometer stretch of beach near the plant where seaweed was found to be 1000 times as radioactive as normal.

Gofman has estimated that the Three Mile Island accident caused at least 333 fatal cancers and leukemias, and his estimate is based upon official dose estimates which many researchers feel dramatically underestimate the actual dose received. Since it is impossible to label a given cancer as having been caused by radiation released at the time of the accident, an exact figure will never be known.

There have been, however, several attempts to analyse disease incidence rates in populations living near nuclear power plants. Dr. Ernest J. Sternglass, a professor of radiological physics at the University of Pittsburgh School of Medicine and an outspoken opponent of nuclear power, has pointed to government health statistics as evidence of the negative health effects of nuclear power plants (*see* Table 15).

TABLE 15

Cancer Death Rate per 100,000 Population

Cancer Mortality Rates in Connecticut and New England Before and After Start-Up of the Millstone Nuclear Plant in Waterford, Connecticut

Site	Approx. Dist. From Millstone	1970	1975	Percent Change
Vermont	200m. NW	176.1	173.9	−1
Connecticut	35m. NW*	168.1	188.4	+12
New Haven, CT	30m. W	200.9	255.5	+27
Waterford, CT	0	152.6	241.8	+58
New London, CT	5m. E	177.4	255.0	+44
Rhode Island	50m. NE	200.1	216.0	+8
Massachusetts	70m. NE	185.0	198.4	+7
New Hampshire	120m. NE	180.4	182.4	+1
Maine	200m. NE	197.7	185.0	−6
U.S.	—	162.0	171.7	+6
New York City	120m. SW	220.9	216.4	−2

*Population center of Connecticut (Hartford-Waterbury area)

Source: Dr. E. J. Sternglass, "Cancer Mortality Changes around Nuclear Facilities in Connecticut," Testimony at the Congressional Seminar on Low-Level Radiation, February 10, 1978, Washington, D.C.

While it is never possible to be sure that such trends are due to the presence of a nuclear plant, the figures do call into question the adequacy of current regulatory radiation release limits.

A similar study of the population living near the Big Rock Point reactor in Charlevoix County, Michigan, found a significant trend toward low-weight births in neighborhoods closer to the reactor. Although the effect may be due to other factors, it would be cavalier to disregard these indications of a potentially serious health effect.

Release Limits:
Where Safety Yields to Expediency
The Case of Tritium

Tritium is a radioactive form of hydrogen which gives off beta radiation as it decays. Produced when water in nuclear power plants is exposed to neutron radiation from the reactor core, this tritiated water is released to the environment routinely and whenever a reactor's cooling system springs a leak. With a half-life of approximately 12.3 years, the substance persists in significant amounts for roughly a century. (The half-life of a radioactive substance is the time it takes for half the amount present to decay. Thus, 1 gram of tritium becomes 1/2 a gram in 12.3 years. The 1/2 gram then decays to 1/4 gram in the next 12.3 year period. And so on . . .)

Tritiated water is particularly dangerous because our bodies cannot distinguish it from normal water. We incorporate it into any part of our body that contains water. Tritium can also cross the placenta from mother to unborn child and be incorporated into the child's body. If the fetus is female and in the process of developing its reproductive organs, the tritium may cause cancer and genetic mutation in the grandchildren of the women who first drank the tritiated water.

Both American and Canadian reactors create tritium.

But Canadian Candu reactors produce more because they use deuterium oxide as cooling water. This type of water, often called "heavy water," contains hydrogen atoms which are one step closer to tritium than common hydrogen. Consequently Candu reactors build up much higher levels of tritium, between 60 and 100 times more than American reactors according to the National Council on Radiological Protection.

Given the ease with which tritium finds its way into our bodies one might expect that allowable release levels would be consistent from country to country, especially when the two countries have reactors on either side of the Great Lakes which provide drinking water to large numbers of citizens on both sides of the border. Such a view would be based on a presumption that public safety or, at least acceptable levels of risk, define the release limits. But unfortunately the amounts deemed acceptable seem to reflect what is easily achievable rather than what is safe. In Canada, where reactors give off at least 100 times more tritium than American ones, the guidelines for water quality allow for 54 times more tritium in drinking water than in the U.S.

New scientific findings indicate that tritium may be more hazardous than once thought due to the unexpectedly long average length of time it stays in our bodies. Despite this new concern, a recent proposal by the Canadian Federal Department of the Environment to tighten water quality standards by a factor of 600 is being met by strong opposition from Canada's nuclear safety regulator, the Atomic Energy Control Board. Though created to insure the public's safety, the AECB has never been indifferent to the health of Canada's nuclear industry.

Radiation from Nuclear Power versus Coal Power

It has sometimes been suggested that when it comes to radiation, coal power is not much better than nuclear power because of the radiation in coal fly ash. While fly ash released from coal plants is a source of radiation which warrants serious consideration, a comparison of radioactive emissions places some perspective on the debate.

According to UNSCEAR data, generating 1 gigawatt of nuclear power will result in a public dose of 5250 person-rems per year while the same output produced with coal leads to a dose of 30 person-rems. And the estimate of a nuclear-related dose doesn't include exposure due to serious nuclear accidents or waste disposal problems.

TABLE 16

Population lung dose-equivalent rate estimates for a
1000 MW coal-fired electric power station

Sector Radius		Assumed Population*	Average Annual Dose Equivalent to Lungs (mrem)†
(miles)	(kilometres)		
0–1	0– 1.61	2,650	0.27
1–2	1.61– 3.22	15,950	3.84
2–3	3.22– 4.83	11,700	2.92
3–4	4.83– 6.44	22,500	1.01
4–5	6.44– 8.05	67,200	0.64
0–5	0– 8.05	120,000	1.35
5–10	8.05–16.09	487,000	0.48
10–20	16.09–32.19	1,233,000	0.11
20–50	32.19–80.47	7,720,000	0.05
0–50	0–80.47	9,560,000	0.1

* *Based on data for the original site for the Newbold Island Nuclear Generating Station (USAEC, 1973).*

† *Calculated on the assumption that the annual dose equivalent rate to the population in all sectors can be calculated using the annual average concentrations predicted by a Gaussian Plume Model (Turner, 1970) with the assumptions of neutral atmospheric stability and windspeed of 6 m/s as annual averages, with uniform wind direction distribution.*

Source: Radiation Exposure Report No. 56, National Council on Radiation Protection and Measurements, 1977 Washington, D.C.

Alternatives to Nuclear Power

Despite the startling numbers which official studies disclose for the health costs of nuclear power, the decline of the nuclear industry has been largely a result of economics. The major competing fuel for electricity generation is coal which is relatively plentiful in North America and, given the costs and economic risks of nuclear power, is more attractive to investors. Of course, there are problems connected with the use of coal: it causes acid rain, is a source of some radiation and poses risks to miners from accidents. But smokestack "scrubbers" are far cheaper and less risky than nuclear plants, and improved safety measures in coal mines would reduce the risks for miners which are no worse than those faced by uranium miners.

Nevertheless, coal is, at best, the lesser of two evils—perhaps reducing the death toll to 50 instead of 1500 per gigawatt year of electricity produced. By far the best alternative is to reduce our demand for electricity by introducing conservation and efficiency measures. "Soft Energy Path" studies in many western countries have shown that even when assuming high rates of economic growth it's actually possible to reduce our reliance on dangerous or expensive fuels by allowing these "softer" renewable technologies to compete on even terms with the "harder" nuclear- and fossil-fuel-based approach.

Given the impact of nuclear power on future generations, the ethical imperative to introduce alternatives is clear. (*See also* Uranium Mines and Mills, Transportation of Radioactive Substances, Radiation Workers, Mining, Radioactive Wastes.)

Uranium Mining and Milling

Uranium mining in North America for both nuclear power production and the nuclear weapons industry has unearthed huge quantities of uranium-rich ore from its natural vault—a vault which would otherwise safely separate this radioactive substance and its radioactive decay products from the human environment. And though neither adequately acknowledged by the nuclear industry nor recognized by the general public, uranium mining creates a serious problem of radioactive pollution.

Only 15 percent of the radioactive substances brought to the surface in a typical uranium mining operation are extracted from the ore; the rest are left in massive tailings piles from which dangerous substances leak into the air and water.

These substances, including uranium, thorium and radium, decay into radioactive gases and dusts including radon and its daughters. According to a U.S. General Accounting Office report:

> Radium is the most significant radioactive waste product in the tailings. It has a very long radioactive life, taking thousands of years before it loses its radioactivity. This loss—called radioactive decay—produces two distinct types of hazards. The first type is highly penetrating gamma radiation. Exposure to sufficient amounts of gamma radiation can cause cancer, such as leukemia. The second hazard, radon gas, produces other radioactive products which attach to particles in the air and are deposited in the lungs when inhaled. Exposure to large concentrations of these radon products can increase the risk of lung cancer.

In the multi-billion year transformation of a uranium atom into non-radioactive lead, radon 222 is the only gas in a long line of solids. It is a short-lived element with a half-life of 3.8 days (compared to 4.5 billion years for uranium 238). When uranium is buried underground in rock, most of the radon created by the natural decay of uranium does not have time to seep into the atmosphere as a gas before it decays into a solid element and is again locked in the ground. As huge amounts of ore are brought to the surface and ground into a powder to enable the removal of the uranium, the ease with which the radon can reach the atmosphere in its few days of life is vastly increased. Once in the air, the radon can be inhaled or it can decay into radioactive daughters which can be washed into the water we drink or breathed in with suspended dust particles in the air. Thus, mining uranium significantly increases our exposure to radioactivity.

With a proper groundcover over a tailings pile the release of radon can be reduced, but engineers are having problems finding an appropriate groundcover—one that both stops radon leakage and won't erode. Erosion is a problem because it uncovers and therefore releases radon and its daughters, and also because radium in various chemical forms is easily washed away. Attempts to stabilize the soil above tailings piles by planting trees have led to the problem of the tree roots providing an escape route for the radon and other radioactive substances dissolved in water.

Leakage of radium has severely contaminated nearby water at a number of uranium mining sites. No fish breed in the Serpent River water system 55 miles downstream from the Elliot Lake uranium mill in Ontario, and the Animus River, which runs near the Durango mill in Colorado, has been severely polluted.

Accidents add to the problem. Several years ago a retaining wall in New Mexico gave way releasing 300 million gallons of radioactive mill wastes into the Rio Puerco River, and more recently a tailings pond overflowed at Key Lake in Saskatchewan, the world's largest uranium mine.

In the past, tailings or refinery wastes were occasionally used as construction site landfill. Houses and schools built over the sites in Colorado and Ontario have been found to contain high levels of radon gas and its daughters (*see* Radon Gas).

Gofman has calculated that radon gas in the vicinity of a tailings pile would give a cigarette-smoking, 50-year-resident male (a person already susceptible to lung cancer) an added lung cancer death risk of four chances in a thousand. Researchers have calculated that in the U.S., a full-scale nuclear power program would create sufficient tailings to cause 450,000 lung cancer deaths in future generations *for each year of mining* due to radon leakage alone. The deaths would occur over tens of thousands of years but the calculation is, nevertheless, a sobering one.

Epidemiological studies of uranium miners who are routinely exposed to radon gas have added yet another disturbing factor to be considered. Radon seems to be more efficient at causing cancer if received in low doses. This is a departure from the linear hypothesis which would suggest the same risk for each added amount of exposure, and a departure which points toward the need for even greater concern about radon exposure. (*See also* Uranium Miners, Nuclear Power, Radon Gas.)

Radioactive Wastes

Every gram of radioactive material created in reactors or mined from deep under the ground presents a waste disposal problem. So far there's no permanent solution.

Low-level wastes are either dispersed in local garbage dumps, (*see*, for example, Smoke Detectors), where they leak into the air and water, or are stored in special dumpsites where an effort is made to isolate them from ground water. Inevitably, however, some of these wastes do reach the water or escape into the air.

High-level wastes, such as those produced by nuclear power and weapons production, are usually kept in temporary storage pools or sit in metal barrels which are slowly corroding.

Despite a serious commitment by government agencies to find a permanent disposal method, no satisfactory method has been developed which can be proven to withstand the ravages of time. The waste problem is one we have created not only for ourselves but also for our descendants.

Indeed, because of the very long half-lives of many highly dangerous radioactive substances, efforts to find a disposal method have focused on burial of high-level wastes in deep holes drilled in what are hoped to be stable rock formations. Ironically, because these stable formations are not very rich in minerals, geologists know relatively little about them.

Unfortunately, the question of what to do with the garbage wasn't answered before we started to create it, and the problem has become an immediate concern.

Hanford, Washington, is the home of one of North America's worst nuclear-waste problems. Approximately 200,000 cubic meters of high-level wastes sit in 150 corroding single-walled tanks. The wastes are largely the result of military fuel reprocessing and officials are now trying to transfer it into newer, double-walled tanks. Low-level wastes on the site (500,000 cubic meters of solid waste and 130,000,000 liters of liquid waste) have already contaminated the soil. Rabbits that inhabit the area leave behind radioactive droppings and some have themselves become radioactive before dying.

The situation at Hanford is, at present, one of the worst cases, but as we produce more and more waste, it may become less of an anomaly.

Of course, few people are happy with the idea of having a nuclear waste dump in their backyards.

In an effort to avoid opposition to radioactive waste dumping near people's homes, some nuclear agencies, including the British, have dumped wastes into the ocean. The practice drew enough opposition to be discontinued, but one British nuclear fuel reprocessing facility at Sellafield continues to pour effluents into the sea. The plant operators claim that most of the radioactive substances have been removed from the water being dumped but scientists in the Canadian Arctic have found traces of radioactive cesium 137 which originates from Sellafield.

Similar discoveries have been made in East Greenland and off the coast of northern Norway.

The irony for the plant operator, British Nuclear Fuels Limited, is that their attempt to avoid opposition at home has now drawn protests from around the world.

In both Canada and the U.S., federal governments have taken on the responsibility of finding disposal sites and a method for disposing of high-level nuclear wastes. In each case local governments may be denied the right to veto a site in their backyard. In

"Long-Term" Radioactive Waste

Dr. Leonard R. Solon, director of the New York City Bureau for Radiation Control, placed the phrase "long-term" radioactive waste in perspective when he noted that " 'Long-term', in this case, dwarfs the span of recorded human history. The half-life of plutonium 239, a significant and inevitable by-product of all nuclear reactors, is 24,000 years. This means that more than 90 percent of the plutonium 239 produced in contemporary U.S. reactors will still be present in 34 centuries, or roughly the span between Amenhotep, pharaoh of Egypt, and Ronald Reagan, president of the U.S."

the U.S., nine possible sites have been selected in six states: Washington, Nevada, Utah, Texas, Mississippi and Louisiana. Canadian efforts are focusing on Central Ontario and Manitoba.

Since local opposition can be assured, we will all likely witness a classic confrontation between the needs of government agencies and the rights of local citizens to a safe environment. The limits to scientific certainty and our obligations to future generations will also come under scrutiny in the search for a disposal method that must stand up for many centuries.

Transportation of Radioactive Substances

The transportation of any hazardous substance poses serious risks to the public. A simple train derailment or traffic accident could result in widespread contamination and radioactive substances are no exception.

The more radioactive the substance being shipped, the higher the risk. For many years small quantities of radioactive material used in manufacturing, hospitals and laboratories have traveled through our cities, usually by road. The radiation given off is usually only of concern to the driver as the substances are kept in shielded containers. However, if a container should be broken in the event of an accident the public as well as police and firefighters are at risk.

The problem is not an insignificant one. The Atomic Energy Control Board of Canada conducted a survey in 1977 and found approximately five hundred thousand shipments of radioactive substances in Canada during that one year. The agency noted that 180 "incidents" came to their attention that year, 20 percent involving releases of radioactive materials into the transporting vehicle or the environment. U.S. numbers are at least ten times higher, and the number of shipments in both countries has grown considerably since then.

Poor warning signs on the vehicles and containers used can add to the risk. In July 1983, a commercial air freight carrier plane crashed in Tennessee. The plane was carrying radioactive material but neither the shipper nor the airplane's operator knew the exact nature or amount. Homes nearby were evacuated until the extent

FIGURE 9

Shipping flasks for radioactive materials are bolted onto a flatbed tractor-trailer.

of the risk could be determined. The incident is an example of the failure of regulators to adequately protect the public from the risks of accidental radiation leakage during transport. Not only was there inadequate labeling, but the first reports of the incident did not include any information about the extent, if any, of the leakage, presumably because local police and fire officials did not have proper equipment on hand to measure radiation emissions.

With the advent of nuclear power has come the need to ship very large quantities of highly radioactive material. At present, many nuclear plants store their radioactive wastes on site, but eventually these wastes will be shipped to long-term disposal sites. The reliability of containers used to ship these extremely toxic substances is a topic of some debate. One U.S. National Research Council study predicts up to 4100 immediate deaths and 680,000 cancer deaths should a load of spent reactor fuel be released by accident in a highly populated area. These figures are based on unrealistically optimistic assumptions— such as a quarter-mile buffer between the accident site and the public, and the ability to evacuate 90 percent of the population in 10-mile radius within four hours. A real accident could be much more serious.

To make matters worse, an accident involving spent reactor fuel may not be the worst case that could arise. In 1980, 105 pounds of plutonium oxide was shipped across the U.S. in a single truck. The substance was being used in a demonstration to test its feasibility as a reactor fuel. Plutonium is one of the most hazardous substances on earth. It has a half-life of 24,400 years and persists in measurable quantities for approximately ten times its half-life, roughly a quarter of a million years. A serious accident involving a large quantity of plutonium could render the accident site and vicinity uninhabitable for centuries and cause thousands of deaths.

The total risk we face cannot be quantified as it depends on the particular circumstances present, but it can certainly be said that the level of risk is increasing as the use of radioctive substances becomes more commonplace.

Nuclear Weapons Production

Most people are well aware of the destructive capacity held by the world's major powers in the form of nuclear weapons. The consequences of a nuclear war would be horrific by anyone's standards, and we need not dwell on them here. What is not as well understood is the cost in human life and suffering due to the existence of these weapons, even if they are never detonated. As well as diverting public resources from social and medical programs into nuclear weapons programs, the production of these weapons exposes the population to lethal radiation.

Most of the exposure is due to testing, but uranium mining, plutonium production and refining, and weapons assembly—all necessary steps in building bombs—are also serious sources of public and occupational exposure (*see* Atomic Bomb Tests).

In 1979, Dr. Carl Johnson released an analysis of cancer rates in the public living downwind from the Rocky Flats nuclear weapons plant near Denver, Colorado. The plant site includes a large cache of plutonium-bearing wastes, some of which have leaked onto the ground where they are picked up by the wind and dispersed widely.

Dr. Johnson's figures, which compared the incidence of various cancers in the exposed public to those in nearby uncontaminated areas, are outlined in Table 17.

TABLE 17

Cancer Rates Among the Public Downwind from the
Rocky Flats Nuclear Weapons Plant

Percentage Increases Above Normal

Type of Cancer	Women	Men
Lung	—	34
Leukemia	—	40
Lymphoma & Myeloma	10	43
Colon	30	43
Ovary	24	—
Testis	—	140
Tongue, Pharynx, Esophagus	100	60

Source: Dr. C. J. Johnson, Radiation and Health, *NIRS*. See Appendix C

Just as in nuclear power production, it is important to remember that much of the public exposure arises from steps in the process that occur outside the nuclear weapons plant (*see* Uranium Mining, Transportation and Radioactive Wastes).

The analogy to nuclear power is also meaningful because the plutonium used in warheads is created in special military nuclear reactors which share many of the safety problems associated with commercial power reactors, including the risk of a major radioactive release due to a serious accident. (In fact, commercial reactors can be used to produce bomb fuel, a serious concern when the export of nuclear power technology is considered.)

Occupational exposure occurs at every stage of weapons production and given the highly refined uranium and plutonium used can be most serious.

But despite the highly toxic nature of these substances, the agencies responsible have been less than successful at keeping track of them. Since 1950, over 4500 kilograms (approximately 9900 lbs.) of enriched uranium have been lost in the U.S., enough to make about 225 bombs of the same size as the one dropped on Hiroshima.

Officials have suggested that the missing uranium 235 has probably been deposited in pipes and filters within the weapons plants and that much of the loss is probably a fictional one due to accounting errors. In light of the potential for damage due to a release into the environment of plutonium or enriched uranium or due to diversion to terrorists, such assurances are small comfort.

The risks of public, occupational and environmental contamination due to weapons production, transport and disposal, while less horrific than those associated with the use of nuclear weapons, should not be overlooked.

Atomic Bomb Tests

Since the explosion of the Hiroshima and Nagasaki bombs, the world has witnessed over 1200 nuclear weapons tests. The legacy of these tests will be many painful cancer deaths as well as children born with genetic mutations, because of the long-lived radioactive substances spread around the world.

Since the 1963 limited test ban treaty, the greatest proportion of testing has been underground, where most of the radiation is contained. But several countries including China, India and France have detonated atmospheric or underwater explosions since that time.

The U.S. stopped atmospheric testing in 1958, but 183 of its 588 tests were above ground and since 1961, 18 of the underground tests have leaked radioactivity into the atmosphere.

The BEIR committee has estimated current exposure rates as equivalent to a whole-body dose of 4.5 millirads per year. The committee has also estimated the dose to sensitive organs that each of us will receive over 50 years due to the various radionuclides.

As with all exposures to ionizing radiation, there is a risk of disease, primarily cancers, but the spectre of genetic damage is of particular concern.

The level of genetic damage is at present unascertainable with estimates of the effect of radiation on the genes varying by a factor of several hundred (*see* Genetic Effects). Easily observed deformities in the first generation of offspring are a tiny fraction of the genetically induced problems which can be expected from radia-

TABLE 18

Fifty-year Dose Commitment from Nuclear Tests
Conducted Before 1971, North Temperate Zone*

Source of Exposure	Dose Commitment, mrads		
	Gonads	Bone-Lining Cells	Bone Marrow
External exposure			
Short-lived radionuclides	65	65	65
Cesium 137	59	59	59
Krypton 85	2×10^{-4}	2×10^{-4}	2×10^{-4}
Internal exposure			
Hydrogen 3	4	4	4
Carbon 14	12	15	12
Iron 55	1	1	0.6
Strontium 90	—	85	62
Cesium 137	26	26	26
Plutonium 239†	—	0.2	—
TOTALS‡	170	260	230

* Data from U.S. Office of Radiation Programs.

† Dose commitment to bone-lining cells has been taken to be equal to integrated dose over 50 yr to bone.

‡ Totals rounded to two significant figures.

Source: BEIR (1980). See Appendix C

tion exposure. The subtle changes in disease resistance which may be a far more common outcome will not be easily identifiable as the result of radiation exposure to a victim's ancestor.

Long-term, multigeneration studies of those living near bomb test sites, where effects are more pronounced, will be needed to estimate the seriousness of the problem. Leukemia incidence among children has already been found to be higher in Utah, close to the U.S. test site, and the lower scores of 17- and 18-year-old students on scholastic aptitude tests has been correlated to the extent of nuclear testing during their early infancy in Nevada and Utah. Nevada and Utah may not be the only areas with particularly high risk. In 1958 a wind shift and temperature inversion trapped a radioactive cloud from weapons testing over Los Angeles for several days. Gofman has estimated that 20,000 Los Angeles residents will die prematurely due to radiation-induced cancer from that one event alone.

Only 475 of the U.S.'s 588 tests were performed in the continental U.S. Most of the remainder were detonated in the Marshall Islands in the south Pacific. Cancer, a disease which was relatively rare there prior to the start of nuclear testing, is now the third leading cause of death among the 33,000 Marshallese. In 1969 the U.S. Atomic Energy Commission declared one of the islands used, Bikini, safe for habitation but a 1977 study found the soil of Bikini Island still contained high levels of strontium-90 and plutonium.

French underwater testing in Polynesia has persisted despite the fears of local residents that the fisheries will be polluted by plutonium. France has, of course, denied that there is a risk but has refused to cooperate in a World Health Organization study of cancer incidence in the Pacific islands. They have also suspended compilation of health statistics in the French Pacific Territories.

The American government has also been guilty of supressing or avoiding rigorous analysis of data on the negative impacts of its bomb testing. Over 250,000 troops were exposed to the various tests conducted in the 1950s and thousands of nonmilitary personnel were also present. The tests also showered nearby residents with radiation. Many of these people have sought compensation from the government and as the evidence comes out, it is becoming clear that the hazards were far greater than were admitted at the time. A rigorous analysis of radiation dose was discouraged by the Atomic Energy Commission. Notes of AEC meetings at the time illustrate the attitude: one AEC commissioner, responding to complaints from Utah residents, is reported as saying "We must not let anything interfere with this series of tests . . . nothing."

More recently the U.S. government has decided not to announce many upcoming underground tests, even though some have leaked radioactivity into the environment.

Radiation from bomb testing is distributed geographically by several mechanisms. The area near a blast (within a few hundred miles) is showered with radioactive dust within the first few hours following a test. Radioactive particles which are thrown into the troposphere descend in rainfalls over several weeks and can travel thousands of miles. Particles that reach the stratosphere (principally strontium 90) are distributed around the world over a period of years.

Once the particles settle on the ground, they are incorporated into the food we eat and the water and milk we drink. Even the Inuit people who live in Canada's north are exposed in this manner, as the caribou they enjoy graze on lichens which have been found to contain high levels of fallout particles. Inuit breast milk has been shown to contain correspondingly high levels of strontium-90.

Strontium 90 is concentrated in the bones and, therefore, raises concerns about both bone cancer and leukemia especially in children as they incorporate more strontium-90 in their growing bones than do adults. Other radionuclides in fallout are iodine 131 which concentrates in the thyroid gland and lesser quantities of carbon 14 which can be incorporated into genetic material and has a very long half-life.

Plutonium was also released by many of the bombs, and it can be expected to cause a great deal of cancer, particularly of the lung. Estimates for the total number of plutonium-induced lung cancers, which have or will result from the tests made so far, are as high as 950,000 deaths located predominantly in the northern hemisphere.

Utah's Mormons are Paying the Price of A-Bomb Tests

Despite the fact that Mormons don't drink, smoke or consume caffeine, if they live in Utah, downwind from the Nevada bomb test site, they suffer an unusually high rate of cancer.

Dr. Carl Johnson found 288 cancers diagnosed between 1967 and 1975 among 4000 Mormon residents of Utah. According to Johnson, that is 109 more than would ordinarily be expected for the group.

CHAPTER FOUR

Radiation
in the Workplace

Prior to the middle of this century, exposure to radiation in the work environment was restricted to a relatively small group of workers, most notably radiologists, pitchblend (uranium) miners and the now famous radium dial painters who licked their brushes which were laden with radioactive radium. These exposures were, in the main, a result of ignorance about the consequences of radiation to human health.

Today the situation has changed dramatically for the better in terms of the degree of exposure an individual worker is likely to receive, but the variety of sources and types of radiation and the number of workers exposed to it has grown steadily.

In 1971 over 800,000 workers in North America were exposed to high-energy radiation as part of their jobs. Current estimates exceed 2 million, and the number is increasing at roughly 10 percent per year.

According to the U.S. Environmental Protection Agency these 2 million jobs are found in a variety of sectors:

Medicine	49%
Industry	19%
Government	17%
Nuclear Fuel Cycle	7%
Other	8%

If we include low-energy radiation exposures from such sources as video display terminals, radio transmitters and microwave food processors, the numbers are growing at so fast a rate that the routine occupational exposure of one person per household is not far off.

Looking only at high-energy radiation exposures, the list of jobs which entail exposure is lengthy:

Hospital and Health Care workers are often exposed to X-ray scatter and to radiation from nuclear medicine treatments to patients. Exposed workers include doctors, nurses, technologists, maintenance crews, laboratory staff and even laundry staff.

Industrial Processes now routinely include X-ray monitoring or radiography for testing welds, metal fatigue, reinforced structures. sealed containers and printed circuits in electronic components. Automatic process controls—which promise to become widely used in the manufacturing sector—often rely on radioactive gauges, which can contain significant quantities of radioactive materials, to regulate thickness, density, rate of flow and level of materials in storage bins.

Electronic and Electrical Technologies often involve high voltages which, as a by-product, give off weak X rays.

Mining has long been known as a high-risk endeavor. Uranium mining is not the only mining process which results in occupational exposure, however. Radon gas is a problem in most hard rock and deep coal mines.

Research and Education exposures are a natural result of the development of processes involving radiation but, many non-radiation researchers also rely on radioactive substances and radiation-emitting devices such as electron microscopes to aid them in their work.

Emergency Service and Security workers such as fire and police personnel will increasingly be exposed to radiation as sources become more prevalent in industry. Radioactive substances which are kept in containers or in restricted access areas can be released into the air by intense heat, explosion or accident during transport. Security workers, particularly at airports, use X rays to scan baggage.

Manufacturing workers can be exposed by processing equipment or by-products. Some product attributes are a result of radiation exposure such as the "grip" of plastic shrink wrap. Manufacturing of products which have radioactive substances within them such as ionizing chamber smoke detectors, static eliminators, luminescent car parts and instrument dials always results in some worker exposure. Radiation-based techniques are replacing many traditional ones to save time or energy. One example is the use of gamma radiation to dry the enamel paint on venetian blinds rather than applying heat.

Nuclear Power and Weapons workers face routine exposure as part of their jobs. Such exposures are not limited to the technicians who operate reactors or assemble weapons. Thousands of truck drivers, security personnel and maintenance personnel also face some exposure.

Low-energy radiation exposure is becoming a fact of life for millions of workers. Those who maintain electric utility lines or radio transmitters, operate radio frequency industrial heaters and sealers, work near radar or microwave ovens, or sit in front of video display terminals, are recipients of radiation in a variety of non-ionizing forms.

Many occupational radiation exposures are at relatively low-dose rates but are of great concern due to the continuous exposure the worker receives and because of the large portion of the population that is exposed. In a world where, before long, every second office worker may sit at a VDT, even a very slight risk to that worker's health or to his or her offspring should not be ignored.

Special Occupational Limits

Government regulators often treat the occupationally exposed in a differing fashion from the general public. Workers whose jobs necessarily involve a likelihood of higher exposure are labeled "radiation" workers and in the case of high-energy radiation are allowed to receive ten times the dose that the general public is permitted. This double standard is defended on several grounds:

i) Unusually disease sensitive individuals have been screened from the workers' group by mandatory preemployment medical examinations, whereas weaker and more susceptible members of the public need the added protection of a stricter limit.

ii) Exposed workers can be easily monitored for the first signs of radiation-induced disease, bettering their chances for survival and, thereby, lowering their mortality risk.

iii) Radiation workers are exposed after informed consent and have, thereby, assumed the added risk voluntarily.

iv) The risk to the exposed worker is said to be roughly equivalent to the occupational risks faced by workers in similar non-radiation related jobs and is, therefore, at an acceptable level.

v) The risk of genetic damage to the population as a whole is unacceptable; the public, therefore, must live with stricter limits.

While it is certainly true that radiation workers may be screened for susceptibility, monitored for disease and have consented to being exposed, the rationales are weak in several respects. Until we know the mechanisms behind radiation-induced disease, screening can only partially insure that susceptible workers will not be exposed. Likewise, monitoring for early signs of disease is not a complete answer given medical science's inability to cure cancer in most cases.

While a worker may be informed enough to consent to exposure and added risk, this is no answer to the problem of genetic damage and risk to future generations. If society feels it must limit the public's exposure to radiation in order to limit genetic damage, on what basis can it approve a higher risk to the health of radiation workers' children?

Finally, the argument that the risk due to the added exposure is in keeping with the level of risk in industry in general is based on misinformation, overly optimistic risk estimates and a philosophy that worker health is an economic value to be traded off against other monetary concerns (*see* Acceptable Risk).

In an ideal world where all risks are known and understood and visited only upon the individuals who accept the risks (and not on their children), and where those individuals freely accept the costs associated with the risk, looser standards could be acceptable. Government regulators, presumably aware that we do not live in such an ideal world, have let groups such as the nuclear power industry persuade them to keep standards loose for economic reasons.

The ALARA Principle

Regulators who set the maximum permissible radiation dose limits for workers usually ask the regulated industry to observe the ALARA principle which calls for doses to be kept "As Low As Reasonably Achievable."

Unfortunately the regulators have not bothered to define the word "reasonable." An employer who has the option of buying radiation shielding and robotics can presumably reduce workers' exposure, if he is willing to pay enough. But when is it reasonable to expect an employer to do so?

Inevitably the "reasonable" limit is one that the employer justifies on economic grounds, weighing the cost of the protective measure against the likelihood of a lawsuit from an injured worker, the possibility of union intervention, and the possibility of the regulator withdrawing an operating license. Worker health becomes an economic issue.

The major problem with this approach is that many of the possible health problems from radiation have no price tag attached to them. If a worker cannot prove that his or her cancer is due to occupational exposure (and this is always a problem for people trying to sue for occupationally related cancers), then the employer does not place much economic weight on that factor; nor are problems which could show up in future generations of economic concern to the employer.

Most often the industry sets a target exposure limit below the regulatory limit. This is done to insure that the regulatory limit is seldom exceeded and not because of the ALARA principle. For example, when asked what the radiation leakage rate was from a food irradiation machine, a company official responded that it was 2.5 millirems per hour. This rate happens to correspond to a dose of exactly 5 rads per year if a worker is exposed 40 hours per week, 50 weeks per year—the maximum permissible occupational dose. He went on to explain that it was unlikely that anyone would spend all their working hours near the unit. He did not, however, even mention ALARA or the possibility of thicker concrete walls to shield workers.

This attitude of treating the maximum permissible limit as an acceptable level of exposure and ignoring the ALARA principle is to be expected in a world where radiation protection is an expensive option. Legal exposure limits should be set with this fact of life in mind. We cannot rely on industry to follow the ALARA principle out of the goodness of their hearts. Sadly, the regulators are not prepared to practice what they preach; industry has demonstrated that it can achieve lower rates than the regulatory limits set.

Radiation Workers

A Public Concern

Any worker whose job creates the possibility of being exposed to ionizing radiation at a level beyond that allowed for the general public is called a "radiation worker." By law radiation workers can receive doses of radiation at a rate of up to ten times the public limit, although their exposures must be monitored.

The issues surrounding the choice of occupational dose limits are not only of concern to radiation workers and their families. We all share the burden of health care costs and lost productivity due to illness. But perhaps a more important point is that the public exposure limit is somewhat arbitrarily set at one tenth the worker limit. Therefore, the factors determining the worker limit indirectly form the basis for the public limit.

It is not unusual for planners of facilities where radiation will be a problem to analyze the safety options available in terms of dollars for each rad of reduced exposure to workers, or more to the point, dollars per life saved. When one considers the direct relationship between allowable public levels and worker levels, and that regulatory limits are currently set at levels felt to be economically achievable, it becomes clear that the public dose limit is to a large extent determined by the economics of nuclear power and other major industrial activities involving radiation. This result has been built into the regulatory system as until recently in both the U.S. and Canada, the regulators setting radiation dose limits were the very agencies responsible for keeping the nuclear industry economically "healthy."

In addition to this trade-off between worker safety and economics, the trade-off between individual and collective risk is also clearly demonstrated in the occupational setting. In an effort to avoid purchasing expensive robotics to reduce worker exposure and to keep individual- and average-worker doses within regulatory limits, the nuclear industry now hires more workers to spread out the exposure. In this fashion, a task which is very hazardous, such as the cleanup operation at Three Mile Island, meets the regulatory limits for worker exposure, but not by limiting the total number of cancers or genetic mutations due to radiation exposure. They simply expose more and more workers. In some cases casual workers are hired for a few days during which they receive the total allowable dose for one year.

Considering the new evidence of the hazards of radiation one might expect both regulators and industry to put limits on the total or collective dose to workers, but the trend has been in the opposite direction.

The approach taken where individual doses are limited but collective dose ignored, leads to the practice of widespread dumping of radioactive substances in the environment. As long as the substances are quickly dispersed in the air or water, no individual gets a large dose. Unfortunately, the population as a whole is likely to suffer the *same* number of cancers and genetic mutations, whether the substances are concentrated in one neighborhood or spread out among many.

Studies of Worker Health

Radiation workers are a logical group to study to determine the health effects of radiation. The workers are exposed to doses which are in the range of common concern (as opposed to the massive doses received by some members of other groups studied, such as the Japanese bomb survivors), and their radiation dose, in recent years at least, is reasonably well monitored.

Results of these studies are of interest to all exposed workers and to the public because they provide useful information about the precise nature of radiation hazards.

Studies have shown, for example, that uranium miners face a high radon gas exposure and suffer a very high lung disease rate. One of the best known studies was conducted by Dr. Joseph Wagoner, who found that rates of malignant and nonmalignant lung disease for miners were 50 to 200 percent above normal.

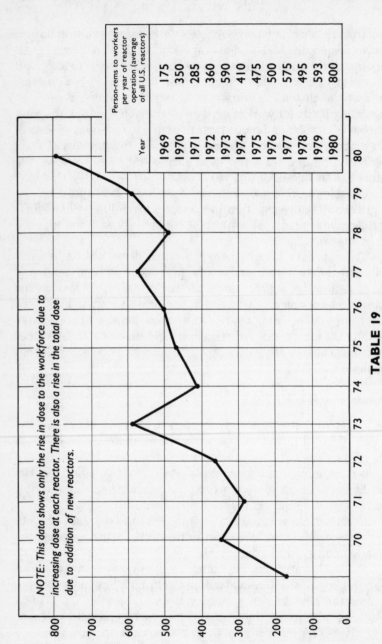

Year	Person-rems to workers per year of reactor operation (average of all U.S. reactors)
1969	175
1970	350
1971	285
1972	360
1973	590
1974	410
1975	475
1976	500
1977	575
1978	495
1979	593
1980	800

NOTE: This data shows only the rise in dose to the workforce due to increasing dose at each reactor. There is also a rise in the total dose due to addition of new reactors.

TABLE 19

Person-rems Per Year of Reactor Operation

Dr. Thomas Mancuso found a significant increase of cancer among nuclear workers at the Hanford, Washington, nuclear facility. Because the exposures received by the workers were well below the regulatory limit of 5 rems per year, the study attracted much criticism from the nuclear industry and federal funding of the study was terminated by the Atomic Energy Commission (the Nuclear Regulatory Commission's predecessor). Funding was later provided by another agency and, despite efforts by some to discredit the study, even critics of the work agree that two forms of cancer, cancer of the bone marrow and cancer of the pancreas, have been shown to occur at elevated levels among radiation workers.

Recent studies conducted by Oak Ridge Associated Universities and the University of South Carolina for the U.S. Department of Energy which analyze worker health at several DOE facilities have found:

• Workers at the Oak Ridge National laboratory have a 49 percent excess leukemia mortality rate compared to the general public. "Leukemia mortality did demonstrate a gradient with increasing radiation dose."

• Janitors, laborers, maintenance men and construction workers at the laboratory have a "significant excess risk" of radiation-associated cancers.

• Workers at Oak Ridge's Y-12 Tennessee Eastman uranium-processing plant between 1943 and 1947 had a "significant excess of deaths from lung cancer when compared to U.S. white male rates."

• Workers at Oak Ridge's Y-12 Union Carbide weapons plant had "excess death for cancer of lung, brain and central nervous system, Hodgkin's disease and other lympathic tissue."

• Workers at the DOE's Fernald, Ohio, uranium-processing plant, have a 36 percent excess of digestive cancers. Also, "there is an association between exposure to uranium and the development of nonmalignant respiratory disease events."

A study of 2529 workers at various DOE facilities who were reported to have received more than 5 rems of radiation in a year found six cases of cancer of the rectum, when only two were expected.

A 1976 study of employees at the DOE's Sevanah river plant (operated by DuPont) was kept secret for seven years until a U.S.

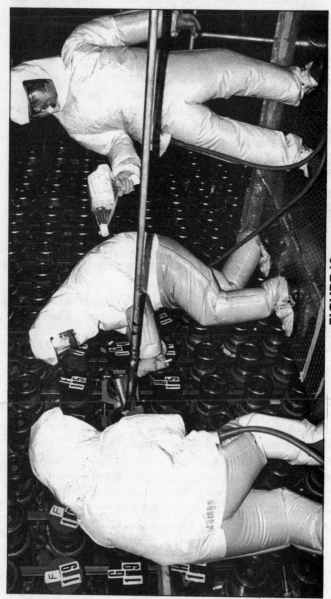

FIGURE 10

Radiation workers at the face of a Candu reactor. Plastic suits reduce exposure due to inhalation, absorption, or swallowing, but do not stop gamma rays.

House of Representatives committee learned of it in 1983. The study found a 60 percent excess of lung cancers in male white-collar workers and a 114 percent excess of leukemia among male blue-collar workers, as compared to DuPont's work force as a whole.

The week before handing over the study, DuPont officials re-analyzed the data in a manner which eliminated the finding of excess cancers. An independent panel of epidemiologists convened by the government's Center for Disease Control in Atlanta reviewed the DuPont study and unanimously condemned the "switch of statistical analysis . . . as inappropriate." They also recommended that the data be reanalyzed by scientists unconnected to the DOE or DuPont.

UNSCEAR (1982) has estimated the radiation dose received by workers in the nuclear power industry at 3000 person-rems per gigawatt year of electricity generated and has estimated 1980 nuclear-power production as approximately 80 gigawatt years. By multiplying the two numbers, we see that 1980 nuclear-power production exposed workers to 240,000 rems. Using UNSCEAR's optimistic total cancer estimate of 100 deaths per million person-rads we can expect about 24 deaths among workers to result from that one year of nuclear power. If we apply Gofman's estimate of 3771 deaths per million person-rads, we can expect 905 cancer deaths.

Nonfatal cancer incidence and genetic damage to future generations must be added to these figures.

UNSCEAR goes on to estimate the world's nuclear capacity in the year 2000 as 1000 gigawatts. If this prediction comes true (which appears highly doubtful in view of the current state of the nuclear industry), we would expect between 300 and 11,300 cancer deaths among workers for each year of nuclear generation depending on which risk estimate is applied. Again nonfatal cancer and genetic damage must be added to the cancer toll.

The above estimates don't include temporary workers, sometimes referred to as "jumpers," who are estimated to collectively receive a dose of radiation about equal to that faced by the entire regular work force. The estimates also exclude occupational exposures which will be faced due to radioactive waste disposal or the eventual dismantlement of nuclear reactors. These factors suggest that a doubling of the estimate is appropriate.

Individual Worker Risk

The risk to an individual exposed worker depends upon the worker's actual exposure, age, sex, health and, in some cases, factors such as cigarette smoking. Table 20 presents estimates of typical individual occupational exposures and associated risks for a variety of jobs.

TABLE 20

Typical Occupation Exposures and Risk*

Job Description	Average dose per year (millirems)	Risk of Radiation-Induced Fatal Cancer
Nuclear Fuel Manufacturing (those exposed to measurable doses)	352.5	3.1%
U.S. (BWR) Reactor Workers	798.5	7.0%
U.S. (PWR) Reactor Workers	725.7	6.4%
Cdn. (HWR) Reactor Workers		
—all	512.5	4.5%
—reactor operations	650.0	5.7%
—mechanical maintenance	1150.0	10.2%
U.S. Medical workers using radionuclide sources	200.0	1.8%
U.S. Radiologists	182.8	1.6%
Cdn. Industrial Radiographers	330.0	2.9%
French Tritium Luminizers	680.0	6.0%
Cdn. Dial Painters	10.0	0.9%
U.S. Dept. of Energy Contractors		
(Academic)	560.0	5.0%
(Research and Dev.)	270.0	2.4%

* *Estimation based upon 30 years of exposure to a male starting at twenty years of age using Gofman's risk estimates and UNSCEAR (1982) dose data.*

The UNSCEAR data for typical exposures are for average exposure. A given individual may face a much higher or much lower actual dose. Anyone who faces such an exposure should insist upon a personal dosimeter which should be worn at all times while at work. Regular readings of the dosimeter will give an indication of the actual dose rate and alert the worker to any unusually high doses or a trend toward increasing doses.

Most occupational settings where radiation is a concern have monitoring programs already in place and have access to experts who can advise on methods to reduce exposures.

TABLE 21

Distribution of Film-Badge Dose Data for
Hospital Radiation Personnel, 1975*

Film-Badge Dose, mrems	Fraction of Personnel, %	Mean Dose, mrems
Nondetectable	43.6	—
100	25.2	41
100–250	12.6	159
250–500	9.0	354
500–750	3.45	618
750–1,000	2.0	867
1,000–2,000	2.53	1,391
2,000–3,000	0.8	2,416
3,000–4,000	0.25	3,391
4,000–5,000	0.19	4,435
5,000–6,000	0.08	5,457
6,000–7,000	0.04	6,500
7,000–8,000	0.03	7,443
8,000–9,000	0	—
9,000–10,000	0	—
10,000–11,000	0	—
11,000–12,000	0	—
12,000–	0.13	128,425

*Data provided by Scientific Committee 45, N.C.R.P., Washington, D.C.
Source: BEIR (1980). See Appendix C

TABLE 22

Person-Rems Accumulated, by U.S. Nuclear Regulatory Licensees, 1973–1976*

Covered Categories of NRC Licensees	Calendar Year	No. Licensees Reporting	No. Persons Monitored	No. Persons with Measurable Exposure	Total No. Person-Rems	Average Exposure per Person (Based on Total Monitored), rems	Average Exposure per Person (Based on Measurable Exposures), rems
Commercial power reactors	1976	62	66,800	36,715	26,555	0.40	0.72
	1975	54	54,763	28,034	21,270	0.39	0.76
	1974	53	62,044	21,904	14,083	0.23	0.64
	1973	41	44,795	16,558	14,337	0.32	0.87
Industrial radiography	1976	321	11,245	6,222	3,629	0.32	0.58
	1975	291	9,178	4,693	2,796	0.30	0.60
	1974	319	8,792	4,943	2,938	0.33	0.59
	1973	341	8,206	5,328	3,354	0.41	0.63
Fuel processing and fabrication	1976	21	11,227	5,285	1,830	0.16	0.35
	1975	23	11,405	5,495	3,125	0.27	0.57
	1974	25	10,921	4,617	2,739	0.25	0.59
	1973	27	10,610	5,056	2,400	0.23	0.47
Processing and distribution of byproduct material	1976	24	3,501	1,976	1,226	0.35	0.62
	1975	19	3,367	1,859	1,188	0.35	0.64
	1974	24	3,340	1,827	1,050	0.31	0.57
	1973	34	4,251	1,925	1,177	0.28	0.61
TOTALS	1976	428	92,773	50,198	33,240	0.36	0.66
	1975	387	78,713	40,081	28,379	0.36	0.71
	1974	421	85,097	33,291	20,810	0.24	0.63
	1973	443	67,862	28,867	21,268	0.31	0.74

*Data from U.S. Nuclear Regulatory Commission.

Source: BEIR (1980). See Appendix C

TABLE 23

Dose Equivalent Received by Transient Workers, 1969–1976* (Jumpers)

	1969	1970	1971	1972	1973	1974	1975	1976
No. workers terminating employment with two or more employers in one quarter	8	29	11	69	157	354	714	1,055
Collective DE, person-rems	5.4	14.6	2.8	61.3	135.5	175.9	507.1	745.3
Average individual DE, rems	0.68	0.50	0.25	0.89	0.86	0.50	0.71	0.71

Data from U.S. Nuclear Regulatory Commission.
Source: BEIR (1980). See Appendix C

Video Display Terminals (VDTs)

Approximately 8 million office workers in North America work in front of a video display terminal (VDT) on a daily basis. The number is rising rapidly as more offices computerize and large numbers of "personal computers" find their way into homes and small businesses. VDTs are quite similar to television screens although they usually include a keyboard and a computer processor.

Fortunately, most screens operate at voltage levels somewhat lower than black and white television sets. Because the voltage is relatively low, in most instances the X rays produced inside the picture tube and from the internal electronic components do not escape in measurable quantities. In fact, a number of studies have concluded that X-ray leakage cannot be detected near properly operating units. Very low levels of weak X-ray leakage are possible, however, if the machine is not working according to its design specifications. This possibility has prompted one official body to recommend legislation requiring testing for X-ray leakage at the time of manufacture and during the life of the unit. At present no such legislation exists.

Color monitors which are sometimes used in office situations may be more hazardous due to their higher operating voltages (see Television Sets). These color monitors are usually used for special graphic applications and should be avoided for routine data and word processing.

Despite test results which disclose little or no X-ray hazards due to VDTs a number of disturbing incidents have spurred continued interest in the health effects of the screens. VDT workers have been found to have several common complaints including eyestrain, blurred vision, fatigue, insomnia, headaches, tension, nausea, irritability, depression and sore necks, backs and legs. Most of these effects are probably non-radiation induced but repeated findings of "clusters" of birth defects and miscarriages among groups of women who have worked near VDTs during their pregnancies point to the possibility of a radiation health problem with the machines.

Only two of these clusters have been investigated using accepted epidemiological study techniques. In both cases the incidence of problem pregnancies was found to be statistically significantly higher in workers exposed to VDTs than in a control group who were not exposed. The studies have not ruled out the possibility of this trend being non-radiation induced. Indeed one Norwegian study has suggested that the electronic components used in some VDTs may leak chemical PCBs, a known carcinogen.

The machines do emit low-energy forms of radiation and this radiation may be the cause of some of the user complaints and may also lead to long-term health problems.

Apart from the visible light coming from the screens, the units give off measurable quantities of extremely low-frequency (ELF) and radio frequency (RF) radiation as well as producing an electrostatic field (static electricity). Ultraviolet and infrared light accompanies the visible light. All of these radiations are produced at levels considered to be well below "acceptable standards" but such assurances are less than comforting given the known or suspected health effects of the various forms.

Lower frequency RF radiation is often referred to as VLF (very low frequency) which covers the frequency range from 3000 to 30,000 Hz (3–30 kHz) or Low Frequency which covers frequencies from 30,000 to 300,000 Hz (30–300 kHz). It is within these ranges that VDTs produce the highest strength "fields." The prime source of the electric field radiation or "E" field within a VDT is the flyback transformer which is commonly located near the rear of the unit. The location of the flyback transformer and the shielding effect of the other components leads to a radiation emission pattern which is not symmetrical. Neighboring workers may even receive a higher dose than the operator of the VDT.

There is also a magnetic VLF field around a VDT which comes largely from the "deflection system" on the back of the picture tube. These low-frequency radiation fields are pulsed fields and as such are more likely to effect our biological functions (*see* Form *versus* Amount in Chapter 7).

Table 24 compares the emission levels of the various types of the radiations found in a Canadian survey of 350 machines with the maximum emission levels allowed or recommended by governments or agencies. The second entry for RF and ELF E-field of 2000 microwatts per square centimeter is the highest rating found in an official U.S. study and illustrates the dramatic difference that can exist between average emissions and the actual emission level of a particularly bad VDT.

While the levels of emission are generally low when compared to the limits it is important to keep in mind three points:

- the limits are often based on inadequate data and are influenced by the needs of large institutions such as the defense establishment (*see* Acceptable Risk);
- VDTs may expose the operator for many hours and over a period of many years, and should be thought of as a "chronic" irritation or stress;
- VDTs emit a wide range of radiations, and it's possible that a combined or synergistic effect may be discovered.

At present there are no large-scale epidemiological studies documenting the exact nature of, and the reasons for the problems experienced by operators of VDTs. Such a study would be expensive but given the widespread use of VDTs, it is clearly justified. Until such a study is conducted a cloud of doubt will remain about the effects of VDTs, especially on fetuses in the womb.

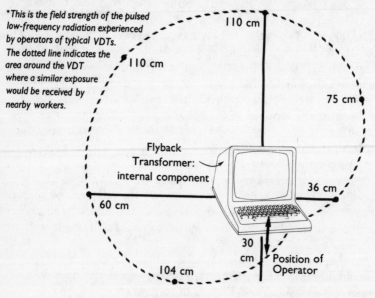

*This is the field strength of the pulsed low-frequency radiation experienced by operators of typical VDTs. The dotted line indicates the area around the VDT where a similar exposure would be received by nearby workers.

FIGURE 11

A Constant Pulsed Field (550 v/m) Surrounding
a Typical VDT*

Source: K. Marha, et al, Publication p83-2E, Canadian Centre for Occupation Health and Safety

TABLE 24

VDT Emissions

Type of Radiation	Typical Emission	Regulatory or Recommended Limit
X ray (mR/h)	0.003	0.5 Cdn. Fed. Reg. U.S. Fed. Reg.
Far Ultraviolet (UV "B" & UV "C") (mW/cm^2)	0.0001	0.1 Cdn. Prov. Rec.
Near Ultraviolet (UV "A") (mW/cm^2)	0.018	0.1 AMA[1] (U.S.) Rec. 0.5 AMA (U.S.) Rec. 1,000 Cdn. Fed. Code
Visible Light (mW/cm^2.sr)	1.4	10,000 Cdn. Prov. Rec. 2,000 U.S. Fed. Reg.
Infrared (mW/cm^2)	0.002	10 Cdn. Prov. Rec. 2,000 U.S. Fed. Reg.
Microwave (E-Field) (uW/cm^2)	0.000007	1 USSR[4] 10 USSR Reg. 1,000 Cdn. Fed. Code 10,000 U.S. Fed. Reg.
RF & ELF (E-Field) (uW/cm^2)	0.013 up to 2000[5]	1[2] USSR[4] 10–750[2] USSR Reg. 1,000 Cdn. Prov. Rec. 10,000 ANSI (U.S.)
RF & ELF (H-Field) (uW/cm^2)	0.03	1[2] USSR[4] 30[3] USSR Reg. 1,000 Cdn. Prov. Rec. 10,000 ANSI (U.S.)

Code = Applicable to specified groups only
Reg. = Regulation
Rec. = Recommendation

[1] *For exposures longer than 7 hours/day*
[2] *Dependent upon frequency*
[3] *Applicable to 30–50 MHz frequencies only*
[4] *For general public (unpublished standard)*
[5] *Highest reading found by U.S. Bureau of Radiological Health (270 v/m measured 30 cm away)*

Preventive Measures

Some employers, including government agencies, have allowed pregnant women to switch to nonVDT jobs during their pregnancy. The approach is a logical one and pregnant women would be prudent to take advantage of the opportunity where it is available. However, such an approach does not help operators who are not pregnant nor does it always prevent exposure of pregnant women in the first weeks of pregnancy, which may be the most critical time.

Other techniques which can reduce exposure include increased distance from the screen, proofreading from a printout rather than the screen when convenient and simply turning the unit off when not in use.

Many newer units have shielding around the flyback transformer or inside the plastic housing or have metal housings which greatly reduce low-energy radiation emissions. When purchasing a VDT, this should be a consideration. A knowledgeable technician can install a grounded metal foil or screen around the sides and back of

Averages Are No Guarantee

The low-average, high-energy radiation emission readings typically found are no guarantee that a particular unit is operating at that level. Numerous reports of clusters of miscarriages prompted the City of Toronto to initiate a review of the known effects of VDTs in 1980. Manufacturers were quick to assure city officials that there was no health risk and that the machines emitted only very low levels of radiation. Three years later city officials were dismayed to learn that 11 of 339 City Hall VDTs checked were found to have "aberrant" radiation levels. One official commented "I didn't think there was a ghost of a chance of finding any high readings after all the assurances from VDT manufacturers . . . this should be a warning to all."

a VDT which will act as a shield without reducing ventilation to the unit or creating a shock hazard.

Recently several manufacturers have started producing VDTs with liquid crystal displays (LCDs) like those used in many digital watches. Use of LCD screens avoids virtually all forms of radiation.

People who use a screen on a regular or prolonged basis should also be concerned with nonradiation-induced problems often referred to as ergonometric issues. Many labor unions and public health agencies are researching and publishing on such topics as adequate levels of lighting and ventilation and appropriate furniture design. For example, it's now known that different colored screens with green or amber backgrounds are easier on the eyes and do not increase radiation levels.

Air Travel

Airplanes themselves do not produce radiation but passengers and crew traveling at high altitudes increase their exposure to natural cosmic radiation (*see* "Natural" Background High-Energy Radiation).

The dose received depends, of course, on the amount of time spent in the air. The International Atomic Energy Agency has estimated that a transatlantic round-trip results in a 4 millirem dose.

UNSCEAR estimates a dose rate of 0.3 millirads per hour of subsonic flight. Gofman has taken the UNSCEAR estimate and his radiation cancer model and calculated the risk for flight crews who are assumed to fly for 1000 hours per year. He estimates added risk to female crew at 1.19 additional fatal cancers per 1000 crew members for each year of employment. The risk for male crew is higher at 1.49 chances per 1000 crew years (*see* Tables 25 and 26).

It is up to the individual to weigh the risks of radiation against the benefits as a consumer or an employee, just as most people do when they compare the risk of airplane crash to the convenience of air travel.

One incidental source of exposure to air travelers is the shipment of radioactive substances on commercial flights. Table 27 gives estimates for this exposure.

TABLE 25

Comparison of Calculated Cosmic-ray Doses
to a Person Flying in Subsonic and Supersonic Aircraft

| Route | Subsonic flight at 11 km | | Supersonic flight at 19 km | |
	Flight duration (hours)	Dose per round trip (millirads)	Flight duration (hours)	Dose per round trip (millirads)
Los Angeles—Paris	11.1	4.8	3.8	3.7
Chicago—Paris	8.3	3.6	2.8	2.6
New York—Paris	7.4	3.1	2.6	2.4
New York—London	7.0	2.9	2.4	2.2
Los Angeles—New York	5.2	1.9	1.9	1.3
Sydney—Acapulco	17.4	4.4	6.2	2.1

Source: UNSCEAR (1982). See Appendix C

TABLE 26

Annual Dose Equivalent from Cosmic Radiation
to Aircraft Passengers and Crew, 1973

| Population Group | No. Exposed | Dose Rates, millirems/year | | Annual Population Dose, person-rems |
		Maximum	Average	
Passengers	35 million	63	2.8	99,000
Cabin attendants	23 thousand	—	160	3,700
Aircraft	17 thousand	—	158	2,650
Total				105,350

Source: BEIR (1980). See Appendix C

TABLE 27

Annual Dose Equivalent from Transport of
Radioactive Materials for the United States, 1975

Population Group	No. Exposed	Dose Rates, mrems/yr		Annual Approximate Collective Dose, person-rems
		Maximum	Average	
Passengers	7 million	108	0.34	2,380
Cabin attendants	40 thousand	13	3	120
Aircraft Crew	30 thousand	2.5	0.53	16
Ground Crew (including bystanders)	720/km²	85		11
Total				2,500

Source: BEIR (1980). See Appendix C

Airport Baggage Scanners

It is now commonplace to be asked to pass your baggage through an X-ray unit before being allowed to board a plane. The X-ray machines used are a source of radiation for the traveling public and the airline attendants who operate them.

The exposure to travelers is relatively low (less than 2 microrems per inspection in the worst case), and the collective dose to travelers is not a matter of serious concern given the obvious benefit of in-flight security. Operators, however, face a relatively continuous exposure to the low levels of X rays leaking out of the units.

Leakage is limited by regulation to less than 0.5 millirads per hour measured 0.5 cm from the unit. If an attendant leaves the unit on continuously and sits or stands in close proximity to it, the dose to the most exposed parts of the body could reach one rad per year—twice the public's upper limit.

Operators should take care to sit away from the units and should consider changing jobs while pregnant.

Uranium Miners

It has long been known that uranium miners suffer from an unusually high rate of lung disease. As early as 1546, pitchblend miners in the Erz Mountains of Central Europe were reported to have a high incidence of what was then known as "fatal lung disease" and in 1879 it was demonstrated that 50 percent of uranium miners were dying of lung cancer in Schneeberg, Germany. The cause of the lung cancer among miners, then and now, is the presence of radon gas and its decay products (or daughters). Radon is one of several radioactive elements in the decay chain of uranium (*see* Radon Gas). Until quite recently the poor ventilation in mines allowed radon gas and its radioactive daughters to build up to dangerously high levels. Today radon gas levels are regulated. Nevertheless, for miners who worked as recently as the 1950–1964 period, their risk of death due to lung cancer is over five times as high as for the general population.

Given current estimates of the lung cancer risk due to radon gas exposure, even the current North American legal dose limit, (which allows 120 WLMs, a dose measure, over a 30-year worklife), allows a legal cancer risk of greater than twice the normal rate.

Recently the Canadian federal regulator of radon exposure, the Atomic Energy Control Board, commissioned a lengthy report to examine the various estimates of risk due to alpha radiation exposure, the predominant form of radiation produced by radon daughters. The report gives a best estimate of the effect of a 50-year occupational exposure to the 4 WLM per year permissible dose of 130 excess lung cancer deaths per 1000 persons, with a possible range of from 60 to 250 deaths per 1000.

For some miners the risks are even higher. Virtually every study of the health risks facing uranium miners points out the dramatically enhanced risks for miners who smoke, a combined risk which not only exceeds the sum of the expected death toll from smoking plus mining but also exceeds the enhanced risk predicted by the relative risk method (*see* Relative Risk).

The best way for a miner to reduce his or her occupational risk is to stop smoking.

Open-pit uranium mining, which is now becoming more common, offers miners the benefits of increased natural ventilation but is not a panacea. Radon gas is quite heavy relative to air and will linger in the pit, keeping levels far above typical atmospheric concentrations. (*See also*: Radon Gas, Nuclear Power, Uranium Mining.)

Non-Uranium Miners

Uranium miners are not the only workers exposed to radon gas underground.

Although the levels of radon in other mines are lower than those experienced in uranium mines, they are nevertheless significant. While the health risk to other miners has not been studied as extensively as the risk to uranium miners, it is reasonable to assume that the risk is at least proportional to the radiation exposure.

An analysis of the average potential alpha energy concentration due to the presence of radon daughters for both uranium and nonuranium mines (including iron, zinc, copper and gold mines) shows that nonuranium mine exposures are between one-half and

FIGURE 12

Workers in uranium mines face high radon-gas levels.

one-quarter that of uranium mine exposures. The data also disclose a progressive trend toward lower levels in both cases presumably due to improved ventilation.

The risk of radiation exposure is present in both hard rock and coal mines. The estimates of lung cancer risk for uranium miners can be divided by four to obtain a rough approximation of the risk faced by other miners (see Uranium Miners).

TABLE 28

Average Potential Alpha Energy Concentration (in WLs) due to Radon Daughters in U.S. Mines

Year	Uranium Mines	NonUranium Mines
1975	0.71	0.31
1976	0.58	0.22
1977	0.51	0.12

Source: UNSCEAR (1982). See Appendix C

CHAPTER FIVE

Radiation

in Medicine

X-rays

Diagnostic X-rays

Of all the radiation exposure that the general public receives, the medical X-ray is by far the most discretionary. It is also one of the few man-made radiation sources that offers a substantial benefit to the recipient. Unfortunately, many people may be getting "too much of a good thing."

Such was the case in the 1970s when routine X-ray breast screening was in vogue. The American Cancer Society and the National Cancer Institute sponsored a program which led to over 250,000 women being X-rayed in order to detect breast cancer. By 1976 several researchers—including some who worked for the sponsoring agencies—were concerned that the program was causing more breast cancer than it was finding. It has been argued that the tests actually produced five cancers for every one found. Since that time the program has been reduced in scope, and exposure levels of the tests themselves have also been cut down, especially for women under 50.

The episode serves as a warning to those who would argue that the diagnostic use of X-rays is of such benefit that it is always worth the risk. But even when the risks associated with diagnostic X-rays are acknowledged by the medical community, the patient is usually kept in the dark, and the doctors themselves are often ill-informed about the extent of the risk.

In many cases, the dose associated with a specific procedure is not accurately known. This leads to great uncertainty about the risk. Moreover, a study conducted in Toronto by Dr. Kenneth Taylor and co-workers found an astounding variability in the dose being given for similar procedures within and between different X-ray facilities in the city. Chest X-ray exposures were found to vary by a factor of six in some cases. Total skin exposures due to "barium meal" examinations of the upper gastrointestinal tract with X-rays and fluoroscopy ranged from 1.6 rads to 90 rads between different facilities (*see* Other Forms of Diagnostic X-rays).

A similar range of from 16 to 128 rads skin dose was found for barium enema examinations of the lower gastrointestinal tract. Within facilities, doses varied by as much as 100 percent for some procedures.

Taylor also examined ways of reducing the dose rate and found that by measuring the output of the equipment, reducing the fluoroscopic exposure rate and using more sensitive film the dose could be dramatically reduced without noticeable changes in the quality of the X-ray films produced. Taylor also points to the need to monitor equipment very regularly to avoid the tendency for exposure rates to "creep upward."

Quite apart from the problems of determining the appropriate dose rate for a given diagnostic procedure, there is the question of the need for such an exposure at all.

Most patients presume that a physician has made a sound decision when he or she orders X-rays, a decision which they don't question for fear of insulting the doctor. Such an approach presumes that the doctor has, in fact, carefully weighed the competing risks and benefits and has placed appropriate weight on the various terms of the equation. But it is often true that the doctor has not bothered to quantify the risk due to the exposure. Patients should insist on some evaluation of the risk and should demand to know all the factors necessary to make an intelligent and informed decision. This may well be an uncomfortable task, but risking one's health is not an appropriate price to pay for a social nicety. Indeed, it should be the doctor who is embarrassed by his or her failure to seek informed consent.

TABLE 29

Typical X-Ray Doses (Millirads)

Type of Examination	Avg. Skin Dose/Film	Avg. No. of Films Per Exam.	Avg. Dose to Bone Marrow/ Examination	Equivalent Whole-Body Dose/Exam	Avg. Gonadal Dose/Exam Male	Female
Mammography (breast screening)	1,500	2/per breast	—	300–600	—	—
Upper Intestine	710	4.3	300	150–400	30	150
Thoracic Spine	980	3	200	150–400	10	10
Lower Intestine	1,320	2.9	600	90–250	200	800
Lumbrosacral Spine	2,180	3.4	200	70–250	1,000	400
Lumbar Spine	1,920	2.9	200	50–180	1,300	800
Intravenous Pyelogram	590	5.3	300	50–150	1,300	800
Cervical Spine	240	3.7	50	40–80	10	10
Cholecystography	620	3.3	100	25–60	5	150
Abdomen	670	1.6	100	10–60	500	500
Skull	330	4	50	20–50	10	10
Lumbo-pelvic	610	1.4	100	5–35	700	250
Chest	44	1.6	40	5–35	10	10
Dental (whole mouth)	910	16	20	10–30	10	10
Hip or Upper Femur (thigh)	560	3	100	2–25	1,200	500
Shoulder	260	2	50	2–25	10	10
Dental (bitewing)	920	3	4	5	2	2
Extremities	100	2.7	10	5	10	10

Source: P.W. Laws, X-Rays: More Harm Than Good? (Emmaus, Pa.: Rodale Press, 1977).

Typical Doses

There are large differences in dose levels between different diagnostic X-ray procedures. Both the energy level of the X-rays and the length of exposure time are important determinants. Of particular importance is the area of the body irradiated. Internal organs, bone marrow and the reproductive organs (gonads) are sensitive targets making chest, head and abdominal X-rays more serious than those of the extremities.

The following chart gives typical doses in millirads for common X-ray procedures. The equivalent whole-body dose takes into account the actual dose to the parts of the body which are exposed, and their relative sensitivity, allowing comparison of cancer risk. The dose to bone marrow is included because of the link between such exposures and leukemia. Gonadal dose (the dose to the ovaries in women and testes in men) shown in the last two columns, is the critical dose when considering possible genetic effects.

The equivalent whole-body dose can be compared to the figures for cancer dose which appear in Table 35 in Chapter 6.

Steps to Reduce Exposure and Risks

- Insist on an explanation of the risk and benefit of all X-rays.
- Take special care when considering exposures to children.
- Avoid exposure if there is any possibility of being pregnant.
- Avoid mobile X-ray units—they tend to give larger doses.
- Ask for a consideration of other tests, (e.g., a skin test for tuberculosis rather than a chest X-ray, physical examination rather than mammography, especially for women under 40).
- Keep track of your X-ray exposures and insist on your X-ray films being shared by doctors and hospitals to reduce the need to make duplicates. Don't be embarrassed about making such a request.
- Avoid fluoroscopy (which is a moving picture form of X-ray), if at all possible, due to the higher dose levels involved.
- Ask for shielding, such as a lead apron, to reduce exposure to sensitive body parts which are not being viewed.
- Where possible, have X-rays taken at a facility where there is a full-time radiologist. (Unnecessarily high exposures are less likely to occur at these facilities.)

- Ask if the facilities have been inspected recently.
- Be as still as possible during the exposure to reduce the chance of a bad exposure and a second one being needed.
- Schedule a physical examination and a dental checkup before a planned pregnancy so that any X-rays required can be obtained before conception.

All of these recommendations must be seen as guidelines only. For example, a particular family history or an emergency may dictate the need for X-rays rather than other diagnostic techniques. To avoid X-rays at all cost is just as misguided as ignoring the risk of unnecessary exposures.

Dental X-rays

While the equivalent whole-body dose from dental X-ray examinations does not come close to the dose due to abdominal X-ray procedures, it is nevertheless a major source of exposure for the average person.

It is difficult to quantify the dose due to a dental X-ray because equipment, film, technique and exposure times vary significantly from one dentist to the next. It is possible, however, to dramatically reduce exposure by using some very simple techniques. One study found that doses to the thyroid could be reduced ten-fold by shielding the cone of the machine which limits the size and shape of the X-ray beam to that of the film being used. Most newer machines incorporate these refinements. Another study found that shielding around the machine combined with a lead apron for the patient could reduce doses to the thyroid by 70.6 percent and to the reproductive organs by 44.5 percent.

If your dentist has a short plastic tapered cone on his X-ray machine, ask him to switch to a long shielded cylinder. This precaution and a reduction in the number of X-rays are probably the best ways to reduce exposure to other parts of the body including the thyroid and brain.

Gofman has calculated the cancer risk for two age groups broken down by cancer type and sex. The figures are based on Gofman's radiation risk estimates, five X-ray films per year (which was the average number of films taken per dental visit in the U.S. in 1972) and an assumed dose per film of 1000 millirads measured at the skin.

TABLE 30

Cancer Deaths due to Five Dental X-rays Per Year
(1000 Millirad Dose to Skin)

Deaths/Million People

	Buccal cavity (mouth and throat)	Brain cancer	Leukemia	Total	Rate (Individuals personal risk)
5 Films/Year from age 7 to 16					
Males	1717	152	22	1891	One in every 529 people
Females	751	122	22	895	One in every 1,117 people
5 Films/Year from age 7 to 45					
Males	3675	316	88	3979	One in every 251 people
Females	1595	260	89	1944	One in every 514 people

Source: adapted from J. W. Grofman, "How Would You Feel About a No-Cost Way to Save a Million Lives?" Committee for Nuclear Responsibility Inc., San Francisco (1982)

Other Forms of Diagnostic X-rays

Fluoroscopy—this is a continuous X-ray. It differs from a regular X-ray in much the same way that a movie differs from a photograph. Radiation doses are much higher in this type of X-ray.

CAT-Scans (Computerized Axial Tomography)—this is a computer controlled X-ray system which takes hundreds of very small "pictures" from different angles around the head or body. The total dose is often comparable to a standard X-ray and the process enables doctors to see very slight differences in the structure of internal tissues (e.g., inside the brain). CAT-scans often make it unnecessary to inject contrast media such as barium or gas, an often painful and dangerous procedure.

Therapeutic X-rays

X-rays are not only used to obtain a picture of the inner parts of the body, they are also used to kill or inhibit the growth of cancer tumors. The X-ray energy and dose for a therapeutic exposure is much higher than for a diagnostic one. Consequently, doctors employ a variety of techniques to focus the dose on the tumor rather than the surrounding healthy tissue, but exposure of healthy tissue is, to some extent, unavoidable.

One interesting feature of X-ray treatment is that some cancers are more sensitive to the effects of the rays at certain X-ray energies than healthy tissue. This difference is exploited by therapists to avoid damage to surrounding tissue when exposing a tumor.

The side effects of X-ray therapy are not just an increased risk of cancer or genetic damage. Often the dose is high enough to cause temporary radiation sickness to the patient. Given the high dose levels involved, doctors have not been able to ignore the risks and use of the technique is usually limited to cases where the risk of no treatment or alternative treatment has been considered and found to be higher than the risks associated with X-ray treatment (*see* Nuclear Medicine, Therapeutic Applications).

Nuclear Medicine

While X-rays have been used as a diagnostic technique and in the treatment of cancer for many years, a number of medical investigative and treatment techniques now employ other forms or sources of high-energy (ionizing) radiation. The practice has become known as nuclear medicine.

Diagnostic Applications

Diagnostic uses of radiation fall into two broad categories: imaging to produce pictures of organs; and radiopharmaceutical studies of body function.

In either case a radioactive substance is introduced into the body either by swallowing it, inhaling it or having it injected. For imaging procedures the patient is then monitored for radioactivity in various parts of the body and a picture of the path of the substance in the body is obtained. This method is sometimes used to find blockages in the lungs or circulatory system or for brain scans used in the detection of a variety of disorders.

Radiopharmaceutical studies test the functioning of various organs and body systems by monitoring the rate at which the radioactive substance is absorbed into the organ or is excreted from the body. For instance, to test the rate at which the thyroid gland soaks up iodine, a radioactive form of iodine is swallowed and then the thyroid gland is monitored for radioactivity.

Typical diagnostic procedures utilizing radioactive substances include studies of:

Thyroid
Liver
Kidney
Mineral metabolism
Localized blood flow
Cardiac system (heart)
Pulmonary system (lung)
Gastrointestinal system (digestive)

Doses vary between different procedures and depend on the particular radioisotope used. Generally the greatest radiation exposure is to the organ being studied but the entire body will often receive some exposure.

As with diagnostic X-rays, the patient should discuss the possible alternatives with the doctor and weigh the benefit against the risk (*see* Diagnostic X-rays). Nursing mothers and those who are pregnant or possibly pregnant should be sure to tell their doctors as the fetus or nursing infant will be exposed. Sometimes tests can be postponed until nursing has been discontinued or until it is certain whether or not you are pregnant.

While use of these procedures has increased over the years, the dose received by the patient has been somewhat reduced by the use of isotopes with short half-lives which decay into nonradioactive elements within hours.

Unfortunately, doctors sometimes fail to properly evaluate the risk of the test in their zeal to diagnose and treat the disease. While the tests are no doubt warranted in a great many cases, it would be unfortunate if the routine use of radioactive substances led to over-use as has been the case with X-rays.

TABLE 31

Estimated Radiation Dose per Diagnostic
Radiopharmaceutical Administration, 1975

Radiopharmaceutical	No. Administrations Covered in Pilot Study	Average Radiation Dose per Administration, mrads		
		Whole Body	Gonad	Bone Marrow
[^{131}I]sodium iodide	814	28	7	12
Other ^{131}I	317	210	204	106
[^{131}I]sodium iodide	326	12	9	10
^{99m}Tc	11,014	177	245	258
^{133}Xe	608	5	5	5
Other	507	1,020	1,020	2,130
Total	13,586	189	242	292

Source: BEIR (1980). See Appendix C

Therapeutic Applications

Therapeutic radiation is most commonly used to treat cancer and to curb hyperthyroidism (an over-active thyroid gland).

Cancer treatment involves either exposure to an external radiation source swallowing or injection of a radioactive substance or surgical implantation of a radiation source in or near the cancer site. The doses given are very high and side effects such as nausea and hair loss are common.

Hyperthyroidism is often treated with iodine 131 which is swallowed and tends to concentrate in the thyroid.

Gofman has reviewed the literature outlining typical doses for iodine-131 treatment. He calculates the cancer death risk *due to treatment* as 5 percent for 25-year-old males treated for hyperthyroidism, and 24 percent for those treated for thyroid cancer. While many researchers would calculate these risks as being substantially lower, the numbers serve to illustrate the importance of a thorough consideration of the competing risks of the treatment and the disease. Some forms of thyroid cancer are less likely to spread or result in death and a proper analysis may find that for a given case the treatment is too risky. The doctor and patient must always weigh the possible effect of the treatment against the known aspects of the disease. Often a very high risk of cancer from treatment which will not likely surface for many years is acceptable given the seriousness of the disease being treated or the advanced age of the patient.

Unfortunately, these treatments sometimes expose persons other than the patient, particularly when the radioactive substance is taken internally or surgically implanted. Patients in adjoining rooms and hospital staff can be exposed to radiation which comes from the patient being treated if the hospital does not take proper precautions.

Accidents do Happen

Nuclear plants are not the only sites where accidents can release radioactive substances. Even medical use of radioactive substances can create serious risks. Recently, a cancer-therapy machine containing thousands of radioactive cobalt-60 pellets was accidently dismantled in the Mexican town of Juarez on the U.S.-Mexico border. Over 7000 pellets fell from the machine as it was loaded onto a pickup truck en route to a metal scrap yard. Five workers received a 400 rem gamma ray dose. Children playing near the pickup, which was parked in a residential area for 50 days giving off 600 rems per hour, were also exposed. Many of the pellets found their way into products made from the refined scrap metal. Radioactive products, such as concrete reinforcing rods and metal table legs, are turning up in a number of locations in the U.S. and Canada.

The example is an unusual one but serves to illustrate that every use of radioactive material is a potential source of accidental exposure.

Ultrasound

In an effort to avoid the known risks of X-rays, doctors are turning to diagnostic techniques using low-energy (non-ionizing) radiation.

Ultrasound is one such technique that bounces pulses of sound at a frequency above the human hearing range off soft tissues within the body. It can be used to identify some circulatory problems but it is most commonly used to view the fetus in the womb. This technique is employed for 60 percent of North American pregnancies to check on fetus development, possible gross deformity and size.

Unfortunately, the use of ultrasound (sonograms) for pregnant women is not above suspicion as a source of risk to the child. One U.S. study of children exposed in the womb has found a trend toward lower birth weight. A British study of childhood cancer found a possibility of a slight association between exposure and development of leukemia and cancer but the findings are very tentative.

Dr. Kenneth Russell, President of the International Federation of Gynecologists and Obstetricians, summed up the situation quite well when he said, "Ultrasound is a form of energy (with the potential to cause damage) and neither I nor anyone else knows for sure what the effects will be somewhere down the road. We don't want another DES situation." (DES was a drug administered to many pregnant women in the 1950s which has now been linked with cancer among the children born.)

Unfortunately, it will be years before doctors know the exact nature and extent of any risk due to the use of ultrasound.

NMR (Nuclear Magnetic Resonance Imaging)

One of the most recent arrivals in the arsenal of diagnostic medicine is a four tonne (8800 lb.), million-dollar device known as an NMR imager. The machine uses an intense magnetic field and a radio frequency signal to create very detailed pictures of the inside of a patient's body.

At present the machines are largely restricted to research applications but are expected to be used in general hospital diagnosis within the next few years. Like ultrasound, the system doesn't involve ionizing X-rays but may, nevertheless, entail some risks due to the strong magnetic field involved (*see* ELF Radiation).

However, the short exposure time may avoid some of the effects which chronic ELF magnetic field exposure is suspected to have.

3

A CLOSER LOOK AT
RADIATION

CHAPTER SIX

High-Energy
Radiation

In Part I we looked at the basic concepts behind current theories on the health effects of radiation. In this part we will examine in greater depth those concepts and the reasons for the high degree of uncertainty in the field. We'll also review risk estimates and regulatory standards which illustrate the differing susceptibilities of different individuals as well as the range of views on the extent of risk.

High-Energy Radiation and Cancer

Cancer is simply the uncontrolled reproduction of cells. Most researchers believe the disease originates in a single cell and is caused by a change within the cell's nucleus (where the genetic material, DNA, resides in the form of genes strung together as chromosomes). While there is no agreement among scientists, the leading theory holds that radiation causes cancer when a unit of radiation energy directly or indirectly disrupts the nucleus destroying the cell's growth and division-control system.

This loss of the cell's ability to control growth may be passed on to each succeeding generation of cells resulting in a mass of such cells, usually taking the form of a tumor. The exact nature of the disruption (and indeed of the cell's regulatory system) is not known. It may involve alteration of the structure of the genes or it may involve a change in the number of genes or the amount of genetic material (the latter likely associated with a disruption occurring during a cell division).

It is presumed that many such disruptions are fatal to the cell. When they are not, it may be many years before the cancer tumor is noticed. In either case, scientists are unable to identify and study that first cancerous cell.

Radiation is not the only mechanism which could induce such disruptions; viruses and toxic chemicals are two other commonly identified suspects. This fact poses problems for radiation researchers for when a cancer is finally observed 5, 20 or 40 years later, it does not have a label telling the doctor whether it resulted from an exposure to radiation, a chemical, or a virus, or if it was due to a spontaneous or "natural" mutation. Furthermore, a disrupted cell in a mother's ovary or father's testes may not cause cancer in the parent but may result in a cancer to the child or a likelihood of cancer among descendants of the child. Finally, irradiation of an embryo or fetus, which most researchers agree increases risk of birth defects, may also induce cancer or a predisposition to cancer in the child.

All of this results in a complex puzzle which has kept scientists from isolating the actual mechanism of cancer induction and radiation's role in it.

Because of this inability to observe the phenomenon thoroughly at every stage in individual humans, science has turned to a different form of detective work, examining the relationship between radiation dose and the incidence of cancer in large populations that have been exposed to radiation. Known as epidemiological studies, examples include studies of survivors of the bombing of Hiroshima and Nagasaki, observers of atomic-bomb tests, radiation workers and groups of medical patients treated with radiation.

Cancer is not the only disease which radiation causes or is suspected of causing, but it appears to be, with one exception, the most important effect on the people exposed. The other primary effect is genetic damage to future generations (*see* Genetic Effects: Our Children's Children's Children).

The Low-Level Dose Response Debate

Arguments about the dangerous effects of radiation exposure often focus on the effects of low doses such as the doses we receive from X rays and nuclear power. We know that very high acute doses of high-energy radiation, such as those experienced near a nuclear explosion, can kill almost instantly. This is due to the mass destruction of the body's cells—destruction at a rate that the body cannot cope with. At medium and medium-low doses or for high doses spread over considerable periods of time the body can cope with the cell destruction although it may result in acute "radiation sickness" for a time. Nausea, hair loss and fever are symptoms of radiation sickness occasionally experienced by cancer victims undergoing radiation therapy. At these dose levels it is also clear that the long-term effects of ionizing radiation include a high risk of cancer and genetic damage. The effects of radiation are not as well understood for the low doses typically experienced in day-to-day living. This is primarily due to the lack of data that would be required to draw clear conclusions.

Fortunately, we are not so barbaric as to expose large populations to high-energy radiation for the purposes of experimentation. We have, however, done so on a number of other grounds, some more acceptable than others. Radiation-exposed medical patients, workers and bomb survivors have provided scientists with the epidemiological data upon which they have tried to postulate models of the relationship between doses of high-energy radiation and diseases such as cancer and leukemia.

The approach is not without its difficulties. Because the exposures were not received in a precisely measured and controlled fashion, there is uncertainty about the extent of the dose. Also the latency period (the time between the exposure and the development of a clinically observable tumor) for cancer is quite lengthy, often several decades, making the measurement of the effect difficult and reducing the value of conclusions drawn before several decades have passed.

Another key problem with the available data is that it is, for the most part, concerned with medium-level radiation doses whereas our greatest concerns center on the low and very low doses typical of routine medical X rays and emissions from appliances, building materials and nuclear power plants. Very few studies have focused on populations exposed to low doses and those that have been carried out have been criticized as not statistically indisputable because the size of the study was not large enough. (Government funding has traditionally been difficult to obtain for large studies that could jeopardize nuclear programs, and to be statistically beyond doubt such studies need to review the radiation exposure and health history of vast numbers of people for several decades.) In the absence of direct studies of low doses of high-energy radiation, much debate has focused on the manner in which the medium dose data can be extrapolated to predict the low-level dose to response relationship.

Researchers disagree both on the interpretation of the imperfect data available and on the manner of extrapolating the conclusions, such as they are, to low and very low doses. The latter argument has led to several schools of thought usually identified by the mathematical formula which defines the shape of the dose response "curve" as it is extended to lower doses. The leading theories are known as the linear, the linear-quadratic, the quadratic and the supra-linear models.

The linear model argues that there is a one-to-one relationship between dose and risk of cancer, i.e., that if *100* units of radiation given to each of 1000 people causes 10 premature deaths due to cancer, then *10* units given to 1000 people will cause one-tenth as many cancers, or one premature death due to cancer. It implies that every dose of radiation is equally dangerous. Most official regulators use this model.

The quadratic model predicts a much lower effect per unit of dose at lower doses of radiation.

And as the name suggests the linear-quadratic approach is a compromise between the two views often defended on the theory that the human cellular repair mechanisms can cope better with the low doses, so we would expect to see a diminished effect per unit of dose at low levels.

Researchers have suggested that for some forms of radiation, especially alpha-particles, research points toward a supralinear curve, especially for leukemia. This would predict that the lower the dose, the higher the disease rate per unit of radiation.

One theory behind this model is that even medium-low doses of radiation kill many cells rather than causing mutation leading to leukemia, whereas very low doses cause serious damage without destroying the cell completely, allowing more potentially cancerous cells to survive. There is some empirical data to suggest such a mechanism but few scientists subscribe to this view at present.

Over the years many researchers and committees have reviewed the available data and expressed a view as to which theory should form the basis for public safety limits. But it would seem that more than just scientific evidence goes into determining which model is supported.

For example, in 1980, the U.S. National Academy of Sciences (NAS) Committee on the Biological Effects of Ionizing Radiation published its third report known as BEIR III. The covering letter from the president of NAS to the Environmental Protection Agency describes the report as having had a "troubled history" and refers to an earlier publicly released version, the distribution of which was suspended. In the earlier report all but 5 of the 17 members of the committee that looked at somatic effects of radiation (effects on the victim rather than genetic effects upon subsequent generations) had agreed to base cancer-risk calculations for low doses on the linear model. That version was revised by six members of the group chosen by the president of the NAS and the new version proposes the linear-quadratic model. The choice of a linear-quadratic model leads to lower estimates of the harm due to radiation exposure from low doses like those the public experienced due to the accident at Three Mile Island. The chairman of the original committee, Dr. Edward P. Radford, has questioned both the validity of the new finding and the process invoked to change the group's earlier report.

Dr. Radford has been reported as saying he suspects the NAS president was pressured to recall the report (which had been released just after the Three Mile Island nuclear plant accident).

Challenging Popular Wisdom

Scientists who have chosen a model which predicts a high cancer-causation effect have, until recently, suffered a good deal of abuse from many of their colleagues. A good example is the case of Dr. Alice Stewart, a British researcher who in 1956 published a paper asserting that a dose of radiation at the level of 1 to 2 percent of what was then commonly thought safe, when delivered to a child in the womb increased the risk of a variety of childhood cancers and of childhood leukemia by about 50 percent. This type of exposure was quite routine in some obstetric practices so it is not surprising that Dr. Stewart's findings were not very well received.

Today there is widespread acceptance of the gist of her findings (although researchers still question her methodology). The case illustrates how the medical and scientific establishment often have a stake in the current wisdom, for they have exposed people to radiation doses based on those standards. A revision toward tighter limits means they have exposed people to greater risk than had been thought, not a particularly pleasant fact to admit. The acceptability of findings such as Dr. Stewart's seems to be changing as new analysis of the data on Japanese bomb survivors has disclosed methodological errors in prior work which had led to a false optimism about the safety of lower-level doses.

Nuclear Plants—Radiation Workers and The Low Level Debate

One of the largest studies of a normal adult population exposed to low levels of high-energy radiation was of radiation workers at the Hanford, Washington, nuclear reactor facilities.

The study by Thomas F. Mancuso, Alice M. Stewart and George W. Kneale looked at 442 radiation workers who died of cancer between 1944 and 1972. The average radiation dose was approximately 1 rem per year. The study found that the risk of fatal cancer was about 8 per one thousand persons exposed to one rem, or one fatal cancer for every 125 person-rems.

This is a risk roughly 20 times higher than previous studies of highly exposed groups had predicted and the controversy following the report's release was considerable.

At about the same time that the study began to show evidence of an increased cancer hazard, the U.S. Department of Energy cut funding to the research. Congressional hearings were held and the National Institute for Occupational Safety and Health (another federal agency) agreed to fund the program. Results of further research have tended to confirm the earlier report.

The nuclear establishment reacted by hiring several groups to refute the Hanford findings. The most thorough of these critics was forced to agree with the trend of the Hanford data for at least two of the three forms of malignancy which were found to increase with radiation dose. The results lend strong support to those who favor a linear relationship between radiation dose and cancer.

This "troubled history" points to the scientific uncertainty and political undercurrent that permeate the study of the health effects of low levels of radiation. The public is left to ponder the conflicting views and interpretations of data, obscured by alarmist headlines in the popular press, scientific jargon, and false assurances by the nuclear industry's public relations teams.

There is no doubt among scientists that there is *some* risk from radiation. When the nuclear industry states there is *no* public health risk following spills of radioactive material they are simply wrong. The question is not whether there is a risk, but *how much* of a risk there is. The second question is to determine if we should err on the side of public safety or public risk when doubt exists. It seems only common sense to avoid situations where there is a reasonable risk of serious harm if there are alternatives, especially if the sacrifice in choosing them is small.

How High-Energy Radiation Interacts With Our Bodies

Radiation is energy. It can be pictured as a minute particle or stream of particles, each particle being a packet of energy. Physicists even consider gamma rays (a form of electromagnetic radiation) to be "particle waves" made up of packets of energy called photons.

When measuring a radiation dose we must consider both the number and nature of these packets and the amount of energy each one carries. It is the energy transferred into the molecules of the cells of our bodies which does the damage. For low-energy (non-ionizing) radiation each packet usually transfers its energy in the form of heat which the cells are able to dissipate without undue damage being done if the rate of bombardment is low enough. But for high-energy (ionizing) radiation the energy level of each packet is high enough that it can disrupt (ionize) the atomic structure by tearing away negatively charged electrons before the cell can dissipate the energy.

These loose electrons and their positively charged counterparts can, in turn, ionize nearby atoms aggravating the negative health impact. Th energy levels involved in radiation-induced ionization are much higher than the energy levels associated with the chemical bonds within proteins and other cellular building materials. One ionization can, therefore, disrupt hundreds or thousands of chemical bonds. It is these disruptions that appear to cause cancer and genetic mutations. The degree of ionization also depends on the electrical charge of the incoming packets and their mass.

The combination of these factors leads to a different pattern of energy transfer and ionization in human tissue for differing forms of radiation. To simplify discussion of the biological damage done by high-energy radiation, two common units of dose measurement are used. One, the rad, is a measure of the amount of energy received by each gram of body tissue. The other, the rem, is similar to the rad but includes a factor which accounts for the added damage large particles or electrically charged forms of radiation can inflict. In many cases the two terms are interchangeable; where they are not, as in the case of alpha particles, rems are the preferred unit.

Exposure to high-energy radiation can be external, due to radiation sources in the environment, workplace or home, or it can

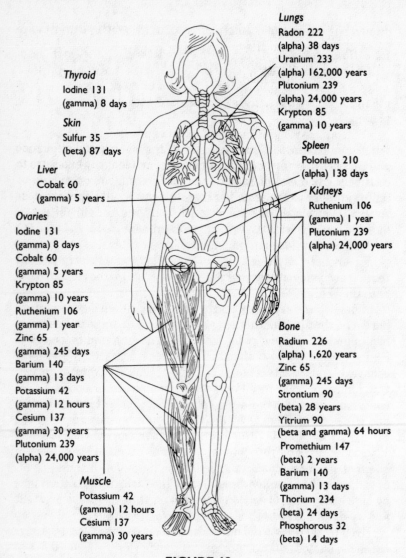

Thyroid
Iodine 131
(gamma) 8 days

Skin
Sulfur 35
(beta) 87 days

Liver
Cobalt 60
(gamma) 5 years

Ovaries
Iodine 131
(gamma) 8 days
Cobalt 60
(gamma) 5 years
Krypton 85
(gamma) 10 years
Ruthenium 106
(gamma) 1 year
Zinc 65
(gamma) 245 days
Barium 140
(gamma) 13 days
Potassium 42
(gamma) 12 hours
Cesium 137
(gamma) 30 years
Plutonium 239
(alpha) 24,000 years

Muscle
Potassium 42
(gamma) 12 hours
Cesium 137
(gamma) 30 years

Lungs
Radon 222
(alpha) 38 days
Uranium 233
(alpha) 162,000 years
Plutonium 239
(alpha) 24,000 years
Krypton 85
(gamma) 10 years

Spleen
Polonium 210
(alpha) 138 days

Kidneys
Ruthenium 106
(gamma) 1 year
Plutonium 239
(alpha) 24,000 years

Bone
Radium 226
(alpha) 1,620 years
Zinc 65
(gamma) 245 days
Strontium 90
(beta) 28 years
Yitrium 90
(beta and gamma) 64 hours
Promethium 147
(beta) 2 years
Barium 140
(gamma) 13 days
Thorium 234
(beta) 24 days
Phosphorous 32
(beta) 14 days

FIGURE 13
High-Energy Radiation and its Concentration in the Body
(*Including half-lives of the radioisotopes*)

Source: Toxicology Mechanisms and Analytical Methods, *Corbet P. Stewart, Academic Press,*
1962

be internal, due to radiation from radioactive substances that have found their way into the body. Different radioactive substances tend to concentrate in, and therefore expose, different parts of the body as shown in the following illustration.

High-Energy Radiation: A Safe Threshold? A Cumulative Effect?

When considering the health effects of high-energy radiation, one of the most important issues is whether or not lengthy exposure to relatively low doses of radiation is less harmful than sudden or acute exposure to the same total dose. Certainly, at very high dose levels we know the body cannot cope with the damaging effects if the dose is received all at once. However, most exposures are in the low or very low dose range.

At the heart of the issue is the mechanism of cellular repair. At the cellular level, it is possible that repair mechanisms can stave off disease if there is a limited "demand" at any given time, but that if a lot of the radiation is received at once, then all the damage cannot be dealt with. This approach assumes there is a difference from the point of view of the cell if the irradiation of that area of the body is fast or slow. The problem with this approach is that the odds are very small that the sensitive area of the nucleus of the cell *is* going to be hit twice. The transfer of energy from a packet of energy is basically an all-or-nothing affair for a given cell, so if the repair mechanism is within the cell, it either copes with the damage or it doesn't. It is possible that neighboring cells or some extracellular repair process can offer "relief," but there is no evidence of such a mechanism.

This leads to the conclusion that each packet of radiation presents the same risk of disease to its recipient regardless of whether it is a part of a flood of radiation or merely a trickle.

The more packets the higher the risk, but for low doses receiving these packets all at once or over time may be equivalent. The total risk is equal to the sum of the risk from each packet. We would, therefore, expect the same risk of cancer from one thousand packets received in one second as one thousand packets received over a year.

This same reasoning along with the data collected in a number of studies would dictate that there is no safe "threshold" dose (for example the study of workers at the Hanford nuclear facility in Washington shows an increased incidence of some forms of cancer among workers who received very low radiation exposures). The effect of one particle or photon on one cell—if it hits the critical part of that cell—is devastating regardless of how many and when other cells get hit.

These conclusions point to the weakness of regulations that limit radiation exposures on a rems-per-year basis since a cumulative dose spread over several years or among many people creates just as much risk of cancer to the exposed group.

A Safe Threshold?

A 1980 joint report from the U.S. Center for Disease Control and the National Institute for Occupational Safety and Health strongly supports the view that very low doses of high-energy radiation are cause for concern. The report is a reanalysis of a controversial 1978 study by Najarian and Colton of Nuclear shipyard workers employed at the Portsmouth Naval Shipyard.

The recorded lifetime occupational radiation exposure among the workers was relatively low, usually less than a doubling of the exposure from natural background radiation. Despite such low dose levels, the study concludes that there is "a statistically significant trend toward increased haematologic (blood) cancer deaths with increased levels of occupational radiation exposures."

This finding supports the view that there is no "safe threshold" of radiation exposure.

From: Theodore Colton, Robert E. Greenberg, John Barron,
Reported in a Memorandum of the Dept. of Health,
Education, and Welfare dated July 9, 1980.

Genetic Effects — Our Children's Children's Children

The nature of virtually every structure and function in a living organism is determined by its genes. Damage to the genes can, therefore, affect every aspect of our health.

The scope of the possible effects of genetic damage is disturbing and so too is the way in which the effects surface, not in the persons exposed to the agent of harm, but in their children or their children's children.

We noted earlier that most researchers agree cancer begins with a single diseased cell. We also noted that the mutation which triggers cancer probably occurs within the nucleus of the cell, where the genetic material DNA resides in the form of genes, contained in structures called chromosomes.

Most cells in a normal healthy person carry a set of 46 chromosomes within which lies the chemically encoded blueprint of our body. The exception to the rule of 46 chromosomes per cell are the sex cells, the male sperm and the female egg (each containing 23 chromosomes), which unite at conception to produce a cell with the full complement of 46 chromosomes. If these sex cells, or the cells they develop from, are damaged, there is a high risk of spontaneous miscarriage or of a child being born with a birth defect.

There is also the risk of damage to the "genetic blueprint" which will not show up in the first generation but may appear in subsequent generations.

This generational separation, coupled with that fact that much of the available data comes from animal studies which are not necessarily indicative of the risk for humans, has left us with a great deal of uncertainty about the extent to which radiation causes genetically induced disease. We do know, however, that the current incidence of genetic disorders, whether caused by radiation or not, is far from insignificant.

The U.S. National Academy of Sciences "BEIR" Committee estimates the incidence of all types of genetic disorders as 107,100 per million babies born. The United Nations Scientific Committee on the Effects of Atomic Radiation (UNSCEAR) estimates 105,000 per million. Dr. John W. Gofman in his analysis of these findings points to the largest component in each estimate, the 90,000 "irregularly inherited" disorders per million, as being an underestimate by a factor of at least three and possibly five. He points out that one disease alone, atherosclerosis of the arteries, is associated with coronary heart disease, cerebral thrombosis, obstructive arterial diseases of the lower extremities and some forms of kidney disease, and that these diseases cause over 50 percent of all premature deaths in the United States.

He goes on to show that atherosclerosis has been associated with a number of genetically inherited traits such as high blood pressure. If half of the cases of atherosclerosis-based disease are inherited due to genetic disorder then roughly 25 percent of all premature deaths in North America may be due to this one manifestation of genetic disorder. It is important to note that genetic disorders leading to a predisposition to disease may play a large role in many other serious health problems, and physicians recognize this every time they ask for a family history.

TABLE 32

Genetic Risk Estimates

Source	Number of Cases Per Million Live Births/ Per RAD/Per Person/Per Generation*
BEIR III (1979)	74–1,132
UNSCEAR (1977)	196
Gofman (1981)	191 to over 20,000

*These figures are for effects after several generations of such exposure.

There is little doubt then that genetic disorders are a serious concern. The question which arises is the extent to which such genetic disruptions are caused by radiation. The estimates are far-ranging. If we look at the effect on a population due to adding one rad of radiation exposure to each of its members in each generation (a dose which is less than half of the dose we continually receive from natural sources of radiation), we can expect anywhere from 74 to over twenty thousand additional genetic disorders per million live births depending on whom we believe. The estimates of the three sources mentioned above are:

Relative Risk

Most of us, after first learning that there are risks attached to something we do, or to an appliance we own, immediately ask the bottom-line question, "How much risk?" How much more likely is it, for example, that a person will get lung cancer if he smokes than if he doesn't?

It turns out that the answers to such questions depend on many different factors including what "population" you belong to in terms of age, sex, geography and occupation. For example, as Table 33 shows, the risk of contracting lung cancer from working in a uranium mine is different for smokers and nonsmokers (700 *versus* 20 deaths per 10,000 per year).

Comparing the risks of a particular radiation dose for different "populations" is a tricky business, largely because of confusion between absolute risk (how many more individuals will get cancer) and relative risk (what the *percentage* increase in cancer incidence will be).

In the field of radiation, risk is most often expressed in relative terms, as the added risk of disease expressed as a percentage of the ordinary risk of contracting that disease without the exposure. Scientists call this approach the relative-risk method. Take the example of two medical patients receiving the same X-ray treatment, one in Country A and one in Country B. Let us suppose we are looking for the risk of cancer caused by the treatment for the two patients. Simply stating that the treatment increases their risk of cancer by, say, 10 percent does not tell us if they face the same risk, because the risk of cancer from all other causes in Country A may be twice that of Country B. Ten *percent* added risk will be twice as much additional risk for the patient in Country A as for the patient in Country B.

As odd as such a finding may seem this is, in fact, just what most scientists suggest is the case. Their explanation goes like this: At least for the risk of cancer and leukemia due to exposure to high-energy radiation, there seem to be differing susceptibilities between different populations and between specific parts of the body. These differing susceptibilities are reflected in the differing "natural" incidences of the disease.

An excellent indication of the validity of this relative-risk approach (rather than simply assuming a given exposure will cause a fixed number of increased cancers per million people exposed) is the case of uranium miners who smoke cigarettes. A 1978 study found that the death rate from lung cancer was somewhat elevated for uranium miners and for smokers but for those miners who also smoked it was extremely high. The actual rates were as follows:

TABLE 33

Lung-Cancer Death Rates Per 100,000 Per Year

Nonsmokers, nonuranium miners	12.5
Nonsmokers, uranium miners	20.0
Heavy smokers, nonuranium miners	265.0
Heavy smokers, uranium miners	700.0

If we looked at the nonsmokers for the *absolute* increase we would find 7.5 extra deaths (20 − 12.5 = 7.5) due to uranium mining. Simply applying this increase to the heavy smokers to find the effect of smoking plus mining we would expect 265.0 plus 7.5 = 272.5 deaths per 100,000 per year, not the 700 actually found. But if we use the relative risk method, we see that the increase for nonsmokers who mine *versus* those who don't is 60 percent. Applying this percentage to the heavy smoker death rate of 265 per 100,000 per year, we would expect 424 deaths per 100,000 per year among heavy-smoking miners. This is still an underestimate, but much closer than the 272.5 predicted by the absolute-risk approach.

Throughout this book the relative-risk method is used and applied to the observed natural incidence of disease in North America.

Those At Special Risk

As we have seen in our discussion of relative risk, certain groups are subject to a greater risk of disease for a given dose of radiation.

The example of cigarette-smoking *versus* nonsmoking uranium miners suggests either a synergistic effect—the sum being greater than its parts when both exposures are simultaneous—or that the exposure to one harmful agent reduces the tolerance to another agent which follows.

In addition to groups who are more sensitive due to exposure to other harmful agents, there is ample evidence of heightened susceptibility for two large groups, the young and the unborn.

Susceptibility of the Unborn. Given the nature of radiation's interaction with living matter at the cellular level, it is not surprising that the unborn suffer a heightened risk.

An embryo develops by repeated division of cells. In a very brief period a single cell becomes the ancestor of millions of cells. A disruption in one cell can, therefore, have much more serious consequences for a rapidly growing fetus than for an adult. Moreover, the fetus's defense systems are not likely to be able to cope due to the number of damaged cells and the undeveloped nature of its defense mechanisms.

As in other areas of study on radiation health effects, there is very little data about the effects on fetuses from human studies, and what information there is has provided the basis for more disagreement than agreement. Despite the problems with the data there is almost universal acceptance of some heightened sensitivity, and medical practitioners routinely avoid exposing pregnant women to X rays.

It seems prudent for pregnant women to extend this caution to all suspected radiation sources if alternatives exist.

Age and Sex. Gofman's analysis of existing data on cancer induction by radiation and its relationship to age and sex are presented in Table 34. The first column refers to the age of the person receiving the radiation. The values in the second column refer to the amount of radiation that will result in one cancer death for that age and sex group. The numbers are for *whole-body exposures* (not localized to one organ) expressed in person-rads. Person-rads is the product of the number of persons exposed times the dose in rads. One person-rad could be the equivalent of one person receiving one rad of radiation or 1000 people receiving one millirad each (a millirad is 1/1000th of a rad). In either case the number of cancers produced in the group or the risk of a cancer being produced is the same for a given age group (according to the linear hypothesis).

If we take 25-year-old males as an example, the cancer dose is 201.4 person-rads. This means that exposing one man to 201.4 rads would in all probability insure he would die of cancer. Alternatively, exposing ten 25-year-old males to 20.14 rads each (one tenth of 201.4), we would still expect one cancer death. Each of the group would suffer a one-in-ten chance of death due to radiation-induced cancer. If twice the radiation dose was given, then we would expect twice as many cancers in the group.

An examination of Table 34 reveals that for any given age and dose, women are less likely to die of radiation-induced cancer (i.e., the dose to insure one death is higher). Also the younger the person or group exposed, the lower the dose required to produce one fatal cancer.

It is important to remember these calculations are for whole-body dose. If the entire dose is concentrated upon one organ, the results will differ depending on the sensitivity of the organ and the natural rate of cancer death due to cancer of that organ.

TABLE 34

Whole-Body Cancer Doses by Age and Sex

Males

Age at Irradiation (years)	Whole-Body Cancer Dose (person-rads per cancer)	Average Loss of Life Expectancy per Cancer (in years lost)
0	63.7	22.3
1	64.5	21.9
2	65.5	21.4
3	67.5	20.9
4	69	20.5
5	70	20.1
6	73	19.6
7	76	19.2
8	80	18.7
9	83	18.2
10	87.8	17.9
11	92	17.4
12	107	17.0
13	123	16.6
14	160	16.2
15	178.1	15.9
16	185	15.6
17	190	15.1
18	194	14.9
19	197	14.6

Age at Irradiation (years)	Whole-Body Cancer Dose (person-rads per cancer)	Average Loss of Life Expectancy per Cancer (in years lost)
20	200.1	14.2
21	200.2	13.9
22	200.4	13.6
23	200.6	13.4
24	200.9	13.1
25	201.4	12.8
26	203	12.6
27	208	12.3
28	214	12.1
29	222	11.8
30	234.2	11.6
31	250	11.4
32	268	11.2
33	288	11.0
34	308	10.8
35	327.6	10.6
36	370	10.4
37	410	10.2
38	450	10.0
39	490	9.8
40	537.5	9.6
41	625	9.4
42	730	9.2
43	860	9.0
44	1020	8.9
45	1232.5	8.7
46	2000	8.6
47	3000	8.4
48	5000	8.3
49	9000	8.1
50	13434	8.0
51	15300	7.8
52	16700	7.7
53	17800	7.5
54	18700	7.4
55	19590	7.1

Continued over

Females

Age at Irradiation (years)	Whole-Body Cancer Dose (person-rads per cancer)	Average Loss of Life Expectancy per Cancer (in years lost)
0	68.3	28.9
1	70	28.4
2	72.5	27.9
3	75	27.3
4	77.5	26.8
5	79.6	26.3
6	83	25.7
7	88	25.2
8	93	24.7
9	98	24.2
10	103.6	23.6
11	115	23.1
12	129	22.5
13	148.3	22.1
14	184	21.5
15	217.2	21.0
16	230	20.5
17	236	20.1
18	241	19.6
19	245	19.1
20	248.5	18.6
21	250	18.2
22	250	17.8
23	250.5	17.3
24	251	17.0
25	251.6	16.6
26	254	16.2
27	260	15.9
28	267	15.5
29	274	15.2
30	284.6	14.8
31	298	14.4
32	315	14.1
33	340	13.8
34	368	13.4

Age at Irradiation (years)	Whole-Body Cancer Dose (person-rads per cancer)	Average Loss of Life Expectancy per Cancer (in years lost)
35	398.7	13.0
36	420	12.7
37	465	12.4
38	510	12.1
39	570	11.8
40	636.1	11.5
41	750	11.2
42	875	11.0
43	1030	10.8
44	1200	10.5
45	1412	10.2
46	2100	10.0
47	4100	9.8
48	7800	9.6
49	11700	9.5
50	14615	9.3
51	16700	9.2
52	18600	9.0
53	19800	8.8
54	20500	8.6
55	20960	8.5

Source: *Gofman (1981)*. See Appendix C

Ionizing Radiation and Aging: The Tri-State Survey

One of the only large-scale nonmilitary controlled epidemiological studies which has considered the health effects of exposure to high-enegy radiation was conducted from 1959 to 1962 in New York State, Maryland and Minnesota. The Tri-State Survey, as it has come to be known, considered a wide range of factors including exposure to diagnostic X rays, and looked at a wide range of health effects.

Key findings include an increased incidence of nonlymphatic leukemia associated with increased exposure to X rays.

Dr. Rosalie Bertell, one of the scientists who worked on the study, in searching for a statistical relationship between leukemia and X-ray exposure, found that the best way of expressing the relationship is to equate one rad of skin exposure and one year of natural aging. She suggests that the bioregulatory mechanism which controls the ability to maintain blood normalcy breaks down with age and exposure to radiation.

Dr. Bertell's theory suggests that cancer and leukemia are not the only bad effects of radiation. Any number of diseases may be due to the breakdown of the body's delicate systems. Indeed, when scientsits have studied groups of patients that have received radiation treatments, they tend to look only for cancer and leukemia as the likely side effects and often presume that other health problems are a result of the disease being treated rather than a side effect of the radiation treatment itself. That presumption is now called into question.

The Tri-State Survey also supported the theory that some people are more vulnerable to radiation than others. Certain common diseases were found to be predictive of leukemia, including asthma, allergies, heart disease, diabetes and certain other bacterial and viral diseases. These health indicators of leukemia susceptibility may simply be evidence of the breakdown of the bioregulatory system, whether due to aging or radiation or some other factor. Thus, further radiation exposure of such persons may be more hazardous than for others. The correlative of this theory is that the breakdown of the system by radiation leaves people more susceptible to these various diseases and possibly a good many others in much the same way as aging does.

Dr. Bertell's statistical methods are innovative

and the scientific community is always wary of
unconventional methods so there is little consensus
about the reliability of her conclusions as yet.

Estimates of Carcinogenicity

Several researchers and agencies have published estimates of the
dose-to-response relationship between high-energy (ionizing) radiation and cancer.

All such estimates must be read in light of the limited data
upon which they are based. To enable comparison, the findings
can be expressed in terms of the number of radiation-induced cancer deaths expected per million person-rads, delivered to a population of mixed ages. A million person-rads is the same as one rad
to each of a million people or half a rad to each of two million
and so on.

TABLE 35

Whole-Body Exposures

Source of Estimate	High-Energy (Ionizing) Radiation-induced-cancer deaths per million persons in a mixed-age population exposed to one rad each[1]
ICRP (1977)	125
BEIR (1979)[2]	177– 353
BEIR (1980)[2,3]	169– 226
UNSCEAR (1977)	75– 175
UNSCEAR (1982)[4]	100– 200
GOFMAN (1981)	3,771
BERTELL (1982)	384–1,450

[1] For gamma, X-ray and beta radiation (effects from alpha or neutrons could be higher per
person-rad).

[2] Relative-risk method.

[3] The BEIR committee of 17 scientists published a volume of results in 1979. In 1980, a final
version rewritten by 6 of the 17, was released with lower estimates; hence, the two differing BEIR
estimates (see pg. 142). The 1980 report includes an estimate of up to 501 deaths per million
person-rads for the relative risk method using the linear model.

[4] UNSCEAR's 1982 report notes that these estimates may go up due to reanalysis of the Japanese
bomb survivor data.

Whole-Body Exposure. The large discrepancy between the different estimates has been discussed in detail by Gofman in his book *Radiation and Human Health*. He has clearly demonstrated deficiencies in the approach of both the BEIR and UNSCEAR committees. His own estimates appear to be far more prudent although his method also displays several weaknesses and has been criticized by a number of commentators.

Gofman's estimate can be expressed in a rough fashion as a four-in-a-thousand chance of death from cancer for the average person exposed to one rad of radiation.

High-Energy Radiation: The BEIR III Report

Dr. Edward P. Radford, who was chairman of the prestigious BEIR III Subcommittee on the Somatic Effects of Ionizing Radiation, but who was excluded from the group that redrafted the report to its final form (based on what appear to be political rather than scientific considerations), has since stated in the December 1981 edition of *Technology Review* "I am convinced that much of the technical information on which the final BEIR III estimates of cancer risk depended is obsolete and that new estimates of the dangers of ionizing radiation are needed. The new evidence indicates that the cancer risks are substantially higher than the BEIR III report concluded."

From: Cancer Risks from Ionizing Radiation by Edward P. Radford, in the November/December 1981 Edition of Technology Review

---

Specific Organ Exposures. The relative-risk approach presumes that the sensitivity of a given organ to radiation-induced cancer bears a relationship to the incidence of cancer of that organ in the population from all causes. Thus, if lung cancer is more prevalent than cancer of the pancreas, we would expect the lungs to be more sensitive to radiation than the pancreas.

Table 36 gives the percentage of cancer deaths due to different types of cancer for men and women.

TABLE 36

Relative Incidence of Fatal Cancer by Organ

Organ or Category	Percentage of Cancer Deaths For Males	For Females
Cancers, overall (excluding leukemia)	100	100
Mouth area	2.9	1.5
Digestive organs	26.5	28.5
Respiratory organs	37.3	15.5
Bone, connective tissue, skin	2.6	2.3
Breasts	0.1	19.9
Genital organs	10.7	12.8
Urinary organs	5.6	3.6
Eyes	0.1	0.1
Brain and central nervous system	2.6	2.5
Endocrine glands	0.3	0.5
Blood and lymph (excluding leukemia)	5.1	5.5
Other	6.2	7.3

Using this data we would expect roughly one-quarter the number of cancers if a radiation dose is to the digestive tract alone as opposed to the same dose to the entire body. Similarly, since there is three times as much endocrine gland cancer as eye cancer among males, we would expect a given dose to the endocrine glands to be three times as dangerous as the same dose to the eyes. By converting various partial body exposures to whole-body equivalents, we can compare the death risk of different types of exposures.

The concept of whole-body equivalent dose or effective dose equivalent (EDE) is used to enable comparisons of cancer risk due to different patterns of exposure. The effective dose equivalent (EDE) is the dose that, if received by the whole body, would result in the same fatal cancer risk as is incurred by the dose actually received by a part of the body. Thus, the EDE for an exposure of the digestive tract to 100 rems would be roughly 26.5 rems because a whole-body exposure of 26.5 rems produces approximately the same risk of fatal cancer overall as a 100-rem dose to the digestive tract.

Of course, this approach only enables us to compare the risk of *death* due to cancer. The type of cancer is determined by the body part exposed, and exposure of some body parts may result in a low EDE because cancer of that organ is not usually fatal. A low EDE tells us that the fatal cancer risk is low, but tells us little about the risk of nonfatal cancer or the risk of other injury such as cataracts.

"Acceptable" Risk and High-Energy Radiation Standards

Numerous official and quasi-official bodies have set acceptable maximum exposure standards for the general population and for workers who face elevated exposure levels.

Most government agencies have relied wholly or in part on the judgment of four of these standards-setting bodies: The National Council on Radiation Protection and Measurements (NCRP), the National Academy of Sciences Committee on the Biological Effects of Ionizing Radiatons (BEIR), the International Commission on Radiological Protection (ICRP) and the United Nations Scientific Committee on the Effects of Atomic Radiation (UNSCEAR).

All of these bodies have principally relied on two sets of human exposure data for their estimates of the risks of radiation: the data from survivors of the Hiroshima and Nagasaki atomic bombs and the data on patients treated for therapeutic reasons, especially those treated with X rays for ankylosing spondylitis, a spinal disorder. Dr. Karl Z. Morgan, a highly respected radiation researcher, has pointed out several of the major drawbacks these sources of data share:

> Should we continue to base our standards on animal studies and on the "accepted" high exposure data from the survivors of the atomic bombings of Hiroshima and Nagasaki and on cancer incidence of ankylosing spondylitis patients who were treated with large doses of x-rays? These Japanese survivors not only received large doses of radiation but also suffered the trauma of blast, fire, deprivation and loss of loved ones, and the ankylosing

spondylitis patients were sick persons who did not live as long as the general population or through the normal cancer incubation period and, hence, as a group had a low incidence of cancer. Both these studies are readily accepted although they contain these serious statistical biases which result in an underestimate of the cancer risk from low level exposure.

Of course, the depleted immune systems of these sick people may also increase susceptibility to cancer, a factor which works in the opposite direction.

The other large source of uncertainty with the data is that associated wth dose estimates. Referring to the ankylosing spondylitis data Dr. John Gofman notes:

> Over a 20-year period, various individuals and groups have wrestled with efforts to assess, in retrospect, what the true dose was to such organs as the lung, lymph nodes, stomach, intestine and pancreas. The dose estimates have varied by factors of 4 or 5 for some of these organs. Obviously, the cancer effects per unit dose will also vary by these factors. The tenuous state of affairs in this important study is best illustrated by the fact that in 1979, some two decades into the study, a new, determined effort is underway to try again better to estimate the dose to various organs.

The Japanese data has long suffered from the uncertainty which flows from reconstruction of exposures after the fact. Location of individuals relative to the bomb blast and their shielding by buildings and clothing must be estimated based on the recollections of the victims. From these estimates scientists have, in turn, tried to estimate the actual radiation dose received.

Recently a more serious flaw has been uncovered in the Japanese bomb dose estimates. Reanalysis of the Hiroshima data indicates that the accepted figures for the amount of neutron radiation released are gross overestimates and the gamma dose may be underestimated. Researchers had attributed much of the cancer incidence at Hiroshima to the very powerful neutron radiation.

The net effect of the new data could be that many of these cancers will be "reassigned" to the gamma radiation-induced category indicating that gamma (and X rays) may be more hazardous than previously thought.

The new dose data is currently being analyzed and research results will become available over the next few years. They may well lead to pressure for a tightening of exposure limits.

If the limits are tightened, it will not be the first time. As scientists have learned more about ionizing radiation, the acceptable dose limit has fallen from 8.8 rems per *day* to 5 rems per *year* for radiation workers and to 0.5 rems per *year* for the general public.

As the tables indicate, workers who are exposed to large amounts of radiation in their occupations are placed in a separate category, that of the "radiation worker," and are allowed to receive ten times the permissible dose for the general public on the theory that the assumption of risk is voluntary and these workers can be monitored for possible health problems (*see* Radiation Workers).

While it is certainly appropriate to protect public health by limiting exposures, the dose-limit approach suffers from a problem of interpretation. The expression "permissible dose" is often assumed to mean a safe dose. There is no such thing as an absolutely safe dose unless the dose is zero.

The question that arises from this observation, and from the acceptance of a linear nonthreshold dose-response model (*see* page 141) is — what's so special about the particular dose limit enunciated by the regulator? The answer appears to be that the limits are politically acceptable in the workplace, i.e., they are easily achieved by industry and those exposed are hard put to fight back because they can't point to many "dead bodies."

One rationale offered to explain the worker limits is that the permissible exposures lead to risks equivalent to those faced by workers in otherwise safe industries. The facts do not support this view. Even if we accept the regulators' analysis of the risk attributable to radiation exposure (which likely underestimates the risk), the limits would allow a radiation worker to suffer a risk 7 to 8 times higher than the average occupational risk in safe North American industries and 2.5 times the risk of workers in the highest risk category—mining and quarrying. In addition, this

TABLE 37

Radiation Worker
Exposure Limits Since 1900

Recommended By	In	Limit
—	1900	10 R*/Day (3,650 R/Year)
Mutscheller & Sievant	1925	52 R*/Year
ICRP	1934	52 R*/Year
NCRP	1934	36 Rems/Year
NCRP	1949	15 Rems/Year
ICRP	1950	15 Rems/Year
ICRP	1956	5 Rems/Year
NCRP	1957	5 Rems/Year

*R = Roentgen = about 0.88 Rems

TABLE 38

Suggested Exposure Limits for the General Public Since 1952

Recommended By	In	Limit
NCRP	1952	1.5 Rems/Year
NCRP	1958	0.5 Rems/Year
ICRP	1959	0.5 Rems/Year
ERDA/NRC	1974	0.005 Rems/Year (for persons living near nuclear plants)
EPA	1977	0.025 Rems/Year (except for thyroid)
EPA	1977	0.075 Rems/Year (for thyroid)

"acceptable" risk is on top of the risk posed to radiation workers by non-radiation occupational hazards.

A second rationalization is that the standards result in an exposure to the average person from nonnatural sources of radiation that is comparable to that received from natural or background radiation. Since background radiation has not been demonstrated to be harmless, the approach is comparable to finding a new pharmaceutical "acceptable" because it merely doubles the natural heart disease rate.

Nature is not always kind and we can do far better than to mimic the less generous aspects of its impact.

The general-public limits seem to be traditionally and arbitrarily set at one-tenth those for radiation workers and, therefore, cannot claim a more acceptable theoretical basis.

Recently both the U.S. Environmental Protection Agency and the Canadian Atomic Energy Control Board, the lead regulators of radiation exposure in their respective countries, have reviewed the radiation standards. Both agencies have proposed new limits which will actually *increase* allowable exposure levels. Labor unions in both countries are opposing the move and are asking for tighter limits than those presently in place.

The most startling feature of the Canadian proposal is that it is based on ICRP risk estimates, the lowest among those of the various agencies and committees usually cited. The other bodies, BEIR and UNSCEAR, offer a range for each risk estimate, whereas ICRP offers one figure near the middle of the UNSCEAR range and toward the lower end of the BEIR range. One would expect that in an area characterized by uncertainty, a public health regulator would wish to err on the side of caution and at least choose the figures in the high range of those offered by the established agencies if not the higher risk estimates offered by many independent researchers.

In the final analysis the only acceptable risk is the risk voluntarily assumed by an individual who has enough information to weigh the risk against the costs and benefits of alternative approaches. Regulators cannot be relied upon to reach the same conclusion.

The ICRP

The International Commission on Radiological Protection (ICRP) has for several decades recommended levels for public and occupational exposure limits for high-energy radiation.

The agency's most recent recommendations in ICRP *Publication No. 26* have caused a great deal of concern among researchers and labor groups. For the first time an ICRP recommendation will, if adopted by national regulatory agencies, result in greatly increased permissible levels of exposure for both the public and workers.

Dr. Rosalie Bertell, a radiation researcher who has spent many years analyzing data on cancer incidence, has little positive to say about the ICRP. Like others, she has pointed out that "Membership in the ICRP is contingent upon nomination by members of the International Congress on Radiobiology and by ICRP members and is subject to approval and selection by the ICRP International Executive Committee. It is a self-perpetuating organization of scientists with a vested interest in the use of radioactive material, not a scientific society based on general professional excellence."

She goes on to point out that the ICRP "has never taken a stand in favor of public or worker health on any major controversial radiation issue" and "has never set up an on-going epidemiological survey, even though its 1959 *Publication No. 2*, clearly stated that this was the only way such effects could be detected." She notes that epidemiologists are excluded from the agency.

Referring to ICRP statements which predict a lesser impact of radiation upon the genes than was once thought, Bertell has pointed to the failure to look for genetic damage among the offspring of nuclear workers and notes simply that "Noncollection of data is not an acceptable method of proof in scientific circles."

CHAPTER SEVEN

Low-Energy
Radiation

Form *versus* Amount

Different types of radiation are categorized by their characteristic
frequencies or wavelengths. As we have seen in the discussion on
high-energy radiation, higher frequency (or shorter wavelength)
corresponds to higher energy. The energy transferred by low-energy
radiation to tissue also depends on the frequency or energy level
of the radiation and on the length of exposure. But the amount of
energy transferred may not be the major determinant of the health
effects suffered. If the mechanism by which low-energy radiation
interacts with our bodies involves our nervous or chemical control
systems, the importance of the frequency and pattern of exposure
may be paramount.

All of us have experienced interference of radio reception
because an electric power tool is being used nearby. The static we
hear is due to the inability of the radio to "tune out" the radiation
coming from the tool through the air or by way of the power lines.
Such interference is often referred to as "noise" by electrical
engineers. The word has a similar connotation to that which we
place on it in everyday life—sound that does not please our ears
because it does not fall neatly into a category like speech or music.
Radios tune in one station and tune out another because they can
zero in on one specific frequency. Interference can be another
signal that coincidentally is at the some frequency, or "noise" that
is at most frequencies and, therefore, inescapable.

Our bodies, like receivers, may not be able to tune out radiation interference with a frequency very close to the internal frequency our bodies use to function, and "noise" that is transmitted from some device across a wide range of frequencies may also present problems.

"Noise" can come from a great many sources. Mathematicians and physicists can prove to us that any single pulse or burst of radiation can be described as equivalent to the effect of superimposing many radiation sources operating at an infinite variety of frequencies. In other words, a single pulse may be seen as a short burst of noise. We can prove this to ourselves by listening to a radio during a lightning storm. Each burst of lightning is a single pulse of radiation, but no matter what frequency we tune our radio to we hear the crackle of the lightning. And just as a single burst of radiation may be experienced as a combination of different frequencies by a radio (or our body), a series of bursts or pulses can also be picked up by radios or nervous systems "tuned" to particular frequencies. Thus, pulsed radiation at virtually any frequency may interfere with our internal body signals.

Examples of pulsed radiation include radar systems and communications systems used for transmitting digitally encoded information, now common in our increasingly computerized world. Examples of single pulses include natural sources such as lightning and artificial ones such as the discharge of electrical fields around high-voltage lines when insulators become moist.

The potential seriousness of the pulsed nature of some radiation is well illustrated in animal studies. When laboratory rats are exposed to strong continuous (nonpulsed) radiation there is no visible effect even after 30 minutes, but when the same strength and frequency of radiation is emitted in a pulsed pattern, the animals will scurry around frantically, begin convulsing within two minutes and die in less than four minutes. Laboratory mice suffer an increased rate of miscarriage and stillbirth when exposed to either pulsed microwave radiation or a pulsed electrostatic field similar to the electrostatic fields that commonly exist between electric appliances such as VDTs and their users. These fields are not, strictly speaking, sources of radiation as no energy moves in the field, but are due to the buildup of electrical charge on the appliance relative to the user or neutral ground. The offspring of mice exposed to the pulsed microwave radiation field were also more likely to suffer low birthweight or deformity of internal organs.

These effects occur in animals at specific power levels and are only suggestive of the risk for men and women. They are, however, strong evidence of the need for caution and further study. They also suggest that the bulk of research which focuses on constant rather than pulsed low-energy radiation may be less than helpful in determining the potential for harm due to low-energy radiation.

Regulatory Standards: A Caution

The fact that most research has focused on nonpulsed radiation is not the only reason why present regulatory standards may be inadequate.

Experiments or calculations which have led to official exposure limits are often based on simplified mathematical models which treat the human body as if it were no more than a column of water. Such an approach can give no insight into the neurological or chemical effects of radiation. Experiments with animals only hint of what we can expect from human exposure. Unfortunately, the largest "experiment" being conducted with humans is the one in which all of us are participants, the flooding of our environment with low-energy radiation from numerous sources. At present, official exposure limits cannot be viewed as safe limits, but only as limits which protect us from only the most obvious and immediate effects of low-energy radiation.

Appendix A

Glossary

Becquerel (Bq)
—one Becquerel equals the amount of a radioactive substance which results in one nuclear disintegration or transformation per second (*see* Curie) (1 Becquerel = 27 picocuries).

Collective Dose
—the sum of all the radiation doses received by all exposed persons due to a given source of radioactivity. (Sometimes restricted to the workforce only.) Usually collective dose is measured in person-rems.

Curie (Ci)
—a term used to describe the amount of radioactive substance present (*see* Becquerel). One curie undergoes 37 billion disintegrations per second. It is the amount of radioactivity present in one gram of radium. (nCi or nanocurie = 1 billionth of a curie; uCi or microcurrie = 1 millionth of a curie; PCi or picocurie = 1 quintillionth of a curie).

Daughter
—the element or isotope to which a given radioisotope decays when it gives off radiation. Daughters can, in turn, decay if they are radioactive themselves.

Decay
—the transformation of an atom to a different form by the release of ionizing radiation.

E-field
—the electric component of microwave, radiowave and ELF radiation. Present whenever there is electric voltage, whether or not there is current flow.

Effective Dose Equivalent (EDE)
—the dose of high-energy (ionizing) radiation to the entire body which would give a fatal cancer risk comparable to that due to the dose to the specific part of the body being considered.

Electromagnetic Radiation
—most radiation is electromagnetic but the term is usually reserved for the low-energy (nonionizing) forms. It is also used to describe X rays and gamma radiation but not for alpha, beta or neutrons.

Epidemiological Study
—a method of studying groups in the population to identify health problems associated with a given factor such as radiation exposure.

Gray
—similar to rad, 1 gray equals 1 joule per kilogram of tissue. Joule is a measure of energy (1 Gray = 100 Rads).

H-field
—the magnetic component of microwave, radiowave and ELF radiation. Present whenever there is electric current flow.

Half-Life
—the length of time required for one-half of the atoms of a radio-active substance to decay to a different form. For example, tritium (a radioisotope of hydrogen) has a half-life of 12.3 years. Thus, if 1 gram of tritium is on hand to start, one-half of a gram will be present in 12.3 years. The important feature of radioactive substances is that this decay is not "linear." Thus, the balance of the tritium does not disappear over the second 12.3-year period, only half of the balance decays, i.e., one-quarter of the original amount over that period. Like cutting a piece of pie repeatedly in half, there is always some left.

Hz (hertz)
—a measure of the frequency of the wave motion associated with radiation (1 Hz = 1 cycle per second).

Ionizing Radiation
—radiation with high enough energy to disrupt atomic structures by separating negatively charged electrons from the rest of the

atom. Ionizing forms of radiation are particularly hazardous to health.

LET (linear energy transfer)
—for ionizing radiation the pattern of ionization in tissues along the path of the ray or particle depends on its energy, mass and electric charge (and to a lesser extent on the density of the tissue). LET is a term used to describe these factors and seems to be important in determining the biological impact of a given form of radiation. The most penetrating forms, X rays and gamma rays, are usually considered low-LET radiation. The least penetrating, alpha radiation, is high-LET radiation and, therefore, more damaging.

Person-Rads
—a measure of the collective dose received by a group of people. It is the average number of rads received by individuals multiplied by the number of exposed individuals. Thus, 10 person-rads can be one person receiving 10 rads or five people receiving 2 rads or 10,000 people receiving one millirad.

Person-Rems
—similar to person-rads but in rems.

RAD
—short for radiation absorbed dose, it describes the dose of ionizing radiation received by living matter in terms of the energy deposited in a given amount of tissue (1 RAD = 100 ergs per gram of tissue). (Erg is a unit of energy.) The rate at which the dose is received is expressed as rads per hour or rads per year (mR (millirad) = 1 thousandth of a rad).

Radioactive Substance
—a substance which gives off ionizing radiation through a natural process known as radioactive decay (eg., uranium, radium, radon gas).

Radioisotope
—most elements (i.e., types of atoms) can be found in a variety of forms. Some elements are unstable and, therefore, radioactive in all or some of these forms. Differing forms are known as isotopes

of the element. Those isotopes that are radioactive are termed radioisotopes.

REM

—differing forms of ionizing radiation will cause differing amounts of damage to human tissue for the same number of rads. Rem is a term used to describe the relative biological impact of the radiation being considered. It is roughly equal to the effect of one rad of gamma rays. Thus a dose of 2 rads of gamma radiation would equal 2 rems, but 2 rads of more damaging alpha radiation may equal 20 rems (mRem (millirem) = 1 thousandth of a rem).

Roentgen (R)

—the earliest measure of the potency of ionizing radiation, it is a measure of the quantity of ionization induced in air. The term is seldom used today (1 roentgen = (approximately) 0.88 rads).

Sievert

—similar to the rem, 1 sievert equals 1 gray of gamma radiation or 100 rems.

WL (Working Level)

—a measure of the rate of high-energy (ionizing) radiation exposure experienced primarily due to alpha radiation in air. (Usually used to measure radon gas exposure in mines.)

WLM (Working Level Month)

—radiation dose due to one working month (170 hours) in a one WL environment.

Appendix B

Who is in Charge and Where to go for Help

In both Canada and the United States the control of radiation-emitting substances and devices is best described as a hodgepodge of regulations and guidelines spread among a variety of departments and levels of government. In some cases exposure is regulated by more than one agency; in others, it falls through the cracks.

The following is an outline of some of the regulated areas and key regulatory agencies responsible for them.

The relevant local agencies (state, province, county and city) vary from one place to the next and are best found by calling the local government switchboard.

Canada

- Radioactive substances and nuclear facilities (including uranium mines): the Atomic Energy Control Board (Federal); note that occupational limits in uranium mines are regulated at both the federal and provincial levels.
- X-ray emitting devices in industrial settings: Provincial ministries of labor
- X-ray emitting devices in medical settings: Provincial ministries of health.
- Radioactive substances in food and water: Federal Department of the Environment, and Provincial ministries of environment, health and agriculture.

Nonionizing Radiation:
  In industrial settings: Provincial ministries of labor (except

for Federal nonmilitary projects: Department of Health and Welfare, Canada, and Federal military: Department of National Defence).

In the environment: Provincial ministries of environment (Federal Department of Environment for federal undertakings). From consumer devices: Federal Department of Health and Welfare (for approval for new devices only).

United States (Federal Agencies)

- Radiation in air and water and primary guidelines for exposure: Environmental Protection Agency.
- Exposure in and near nuclear facilities: Nuclear Regulatory Commission.
- Occupational exposures (exposures not otherwise federally regulated): Occupational Safety and Health Administration (OSHA, Federal Department of Labor).
- Mining exposures: Mine Safety and Health Administration (Federal Department of Labor).
- Transportation of radioactive substances: Federal Department of Transportation and the U.S. Postal Service.
- Consumer products: Consumer Products Safety Commission and the Food and Drug Administration.
- X ray, radiation-emitting machines and food and drugs: Food and Drug Administration.
- Radiation in government-owned facilities and Federal employee exposures: Department of Defence, Department of Energy, Veterans' Administration, Department of Health, Education and Welfare.

Appendix C

Further Information

Ionizing Radiation

Gofman, John W., M.D. *Radiation and Human Health*. San Francisco, California: Sierra Club Books, 1981.

 This is an excellent text which provides a methodology for calculation of risk due to a given exposure. Gofman spells out all his assumptions and calculations and offers a useful critique of the methodologies employed by UNSCEAR, BEIR and others. (Technical and readable.)

Bertell, Rosalie. *Handbook for Estimating Health Effects from Exposure to Ionizing Radiation*. Buffalo, N. Y.: Ministry of Concern for Public Health, 5495 Main Street, Suite 147, Buffalo, N.Y., 14221, August 1984.

 Compiled for the West German Parliament, this technical guide is useful for converting known radiation doses to estimated health effects.

UNSCEAR, (United Nations Scientific Committee on the Effects of Atomic Radiation). *Report of the United Nations Scientific Committee on the Effects of Atomic Radiation to the General Assembly.* New York: United Nations Publications (Sales No. E.82.IX.8.), 1982 (with annexes).

 A good source of information on the actual exposures faced by the public and workers due to the various uses of ionizing radiation around the world. (A data source above all else.)

BEIR, (Committee on the Biological Effects of Ionizing Radiations). *The Effects on Populations of Exposure to Low Levels of Ionizing*

Radiation. Washington, D.C.: National Academy Press, National Academy of Sciences, 1980.

Like UNSCEAR's report, this is one of the fundamental references in the field. Unfortunately, the methodology and assumptions behind its conclusions are not clear, and it may be a better example of how science clashes with politics than anything else (*see* The Low-Level Dose-Response Debate). It is, nevertheless, a useful source of information on radiation exposures in the United States. (Technical data.)

Nuclear Information and Resource Service. *NIRS Radiation Packet.* Washington, D.C.: Nuclear Information and Resource Service, 1536 Sixteenth Street, N.W., Washington, D.C., 20036 (updated frequently).

NIRS is an excellent source of information on nuclear issues in general, and their packet of information on radiation contains a good selection of popular and scientific articles reprinted from diverse sources. (An excellent source of data and analysis, most of which is quite readable.)

Laws, Priscilla W. and the Public Health Research Group. *The X-ray Information Book: A Consumer's Guide to Avoiding Unnecessary Medical and Dental X-Rays.* New York: Farrar, Straus and Giroux, 1983.

Adapted from Laws' earlier book, *X-rays: More Harm Than Good?* The titles say it all. (Easy reading.)

Non-ionizing Radiation

Becker, Robert O. and Andrew A. Marino. *Electromagnetism and Life.* Albany, N.Y.: State University of New York Press, 1982.

This book summarizes much of the research and current thought of the interaction between low-energy (non-ionizing) forms of radiation and biological systems. (Technical and readable.)

Assenheim, H.M., A.B. Cairne, D.A. Hill and E. Preston. *The Biological Effects of Radio Frequency and Microwave Radiation.* Ottawa, Ont.: National Research Council of Canada (Publication No. 16448), 1979.

A useful technical reference on the subject which summarizes research in the area up until 1979. (Technical data source but, nevertheless, understandable.)

Brodeur, Paul. *The Zapping of America.* Don Mills, Ont.: G.M. McLeod, 1977.

This book, which was excerpted in the December 1976 issue of *The New Yorker* magazine, gives a wonderful account of the scientific and political battle fought over microwaves in the U.S. up until 1976. (A good read.)

Ott, John. *Health and Light.* New York: Simon and Shuster Pocket Books, 1977.

Ott is a somewhat controversial researcher with an important message about the effects of nonnatural lighting. (Technical and readable.)

Radiation Alert Detector, a portable detector that gives a continuous meter readout of the ambient radiation level (alpha, beta, gamma and X ray) up to 2,000 times normal background in milliRoentgens per hour (mR/hr); warning lights and tones are activated when the radiation level exceeds normal background. (Normal background is about 0.01 mR/hr.) Radiation Alert (approx. $280) and its smaller version, Radiation Alert-Mini (approx. $200) are available from The Natural Rights Center, The Farm, Summertown, Tennessee, 38483.

Index

About Energy Probe

One of the most effective environmental organizations on the continent, Energy Probe first made headlines coast to coast in 1975 with its exposé of radioactive contamination in the town of Port Hope, Ontario. The national scandal forced the Canadian government to clean up the hazard.

Since then the organization has been active at the local, national and international level on critical questions of environmental energy policy and public process.

Whether Energy Probe is seen as an environmental protection group, a consumer advocacy organization or simply a citizens' watchdog, its independence from government funding and vested interests has led to widespread acceptance of its reports and studies and has won the grudging respect of its traditional adversaries.

Energy Probe is a non-profit registered charity. Inquiries and donations can be directed to:

Energy Probe
100 College Street
Toronto, Canada
M5G 1L5